OXFORD MEDICAL PUBLICATIONS

In the Shadow of the City

In the Shadow of the City

Community Health and the Urban Poor

Edited by

TRUDY HARPHAM

Evaluation and Planning Centre for Health Care,
London School of Hygiene and Tropical Medicine

TIM LUSTY

Health Unit, OXFAM UK

and

PATRICK VAUGHAN

Evaluation and Planning Centre for Health Care,
London School of Hygiene and Tropical Medicine

OXFORD   NEW YORK   DELHI
OXFORD UNIVERSITY PRESS
1988

Oxford University Press, Walton Street, Oxford OX2 6DP
Oxford New York Toronto
Delhi Bombay Calcutta Madras Karachi
Petaling Jaya Singapore Hong Kong Tokyo
Nairobi Dar es Salaam Cape Town
Melbourne Auckland
and associated companies in
Beirut Berlin Ibadan Nicosia

Oxford is a trade mark of Oxford University Press

Published in the United States
by Oxford University Press, New York

British Library Cataloguing in Publication Data
Harpham, Trudy
In the shadow of the city: community
health and the urban poor.
1. Community health services—
Developing countries
I. Title II. Lusty, Tim III. Vaughan,
Patrick, 1937–
362.1'0425   RA441.5
ISBN 0-19-261591-2
ISBN 0-19-261698-6 Pbk

Library of Congress Cataloging in Publication Data
In the shadow of the city: community health and the urban poor/
edited by Trudy Harpham, Tim Lusty, and Patrick Vaughan.
p. cm. —(Oxford medical publications)
Based on an international workshop held at Oxford, United Kingdom, July 1985.
Bibliography: p. 213 Includes indexes.
1. Urban health—Developing countries—Congresses. 2. Urban poor—Medical care—
Developing countries—Congresses. 3. Community health services—Developing countries
—Congresses. I. Harpham, Trudy. II. Lusty, Tim. III. Vaughan, Patrick.
IV. Series.
[DNLM: 1. Community Health Services—organization & administration—Congresses.
2. Developing Countries—Congresses.
3. Health—Congresses. 4. Poverty—Congresses. 5. Urban Population—Congresses.
WA 395.I35 1985]
RA566.5.D44I5 1988 362.1'0425–dc19 87–28226
ISBN 0-19-261591-2
ISBN 0-19-261698-6 (pbk.)

Typeset by Colset Private Limited, Singapore
Printed in Great Britain by
St Edmundsbury Press Limited
Bury St Edmunds, Suffolk

Foreword

H. MAHLER, MD

Director-General, World Health Organization

National and international awareness of the plight of underserved urban and peri-urban areas is increasing, especially in the developing countries. Although urban health services absorb the majority of national resources for health, the great majority of the underprivileged in cities still have limited access to health care.

This book reviews the health and health-related problems and needs of unserved and underserved population in urban setting; it illustrates selected experiences of governments and non-governmental organizations which face these major and ever-increasing problems and are determined to bring about solutions.

I consider that within the diversity of these experiences there are some common messages to improve the quality of life of the underprivileged populations. We should review these experiences, add to our own knowledge, and then determine our new approaches and action in the health care of urban populations.

Dedication

This book is dedicated to all people who live
and work in the slums, shanty towns and
squatter settlements in the
developing world.

Preface

The health of people in the slums, shanty towns, and squatter settlements of developing countries is a neglected issue and until recently there was very little international attention focused upon the health of the urban poor. Gradually, more people are appreciating the extent and the growing nature of the problem. Since 1983, a number of international agencies have begun to consider the issues associated with urban primary health care.

In July 1985 an international workshop on Community health and the urban poor was held at Oxford, United Kingdom. The meeting was funded jointly by UNICEF, OXFAM UK, the London School of Hygiene and Tropical Medicine and the Overseas Development Administration, UK. Fifty participants representing 30 countries prepared papers which addressed issues concerning health and the urban poor in developing countries, or described case studies from poor urban communities. This book is a result of that workshop and it is intended for those interested in urbanization, the condition of the urban poor and primary health care.

As this book is part of a vanguard focusing upon health problems and programmes in poor urban communities, the perspective is global with case studies from a variety of countries. Certain characteristics and health problems of poor urban communities differ between regions and countries, but it is not our intention to take a specialized, regional perspective. It is hoped that this book will stimulate such work in the future.

It would be very difficult, and it has not been our intention, to cover all aspects of urban health care in one book. Two subjects which are not covered in depth åre the specialized subject of occupational health and the costs and financing of urban primary health care, for which there is very little data available.

Many individuals and communities in poor urban areas have long since recognized the need to work towards better health and are doing something about it. We hope that this book amplifies their voices and encourages more international action and interest in the health of the urban poor.

T.H.
T.L.
P.V.

London 1987

Acknowledgements

PART EDITORS: Bill Cousins, UNICEF (Ch. 14) and Brian Pratt, OXFAM UK (Ch. 7).

EDITORIAL ASSISTANTS: Charlotte Dorell and Gita Patel (Middlesex Polytechnic, London).

Special thanks to all the participants at the workshop on 'Community Health and the Urban Poor' held at Oxford in 1985 who are acknowledged individually in the list of contributors. Colleagues in the Evaluation and Planning Centre, London School of Hygiene and Tropical Medicine, OXFAM UK, and UNICEF provided much stimulus and support, particularly David Ross, Susan Rifkin, and John Donohue. This book would not have been possible without the workshop at Oxford which was organized so well by numerous people, including Brian Cockrell, Louise Cross, Diana Hasting, and Joan Turner. Photographs were kindly provided by Vinod Virkud (who took on a special assignment in Bombay), Anne Charnock, Catherine Goyder, Arif Hasan, Lewnidah Ongari, UNICEF, WHO, OXFAM UK, Christian Aid, and Earthscan. Doug Brunner let us use his cartoons freely, and Teaching Aids at Low Cost (TALC) provided diagrams. No acknowledgements would be complete without mention of Alessandro Rossi-Espagnet. For years a voice in the wilderness, Alessandro has been calling for more attention to urban primary health care, until his recent retirement from WHO. This book builds upon the foundations laid by him, and the papers and comments provided by the workshop participants.

We are grateful to the Overseas Development Administration, UK, UNICEF and OXFAM UK who provided a subsidy to enable this book to be published at relatively low cost.

Contents

Contributors

The following people prepared papers for the workshop on Community Health and the Urban Poor, held in Oxford, United Kingdom, for one week in July 1985:

Jane Banez, PHC for Asia, World YWCA, Quezon City, Philippines

Vesna Bosnjak, UNICEF, Mexico City, Mexico

Reginald Boulos, Complexe Medico-Social de la Cité Soleil, Port Au Prince, Haiti

Ruth Brown, Operation Friendship, Kingston, Jamaica

Sandy Cairncross, London School of Hygiene and Tropical Medicine, UK

Celerino Carriconde, Casa Amarela Community Health Project, Recife, Brazil

Edgardo Cayon, UNICEF, Mexico City, Mexico

Samir Chaudhuri, Child in Need Institute, Calcutta, India

Victor Choquehuanca-Vilca, Municipality of Lima, Peru

Mac Corry, Order of Saint John, London, UK

Hilary Creed-Kanashiro, Programa de Salud, Pamplona Alta, Lima, Peru

Ralph Diaz, UNICEF, Nairobi, Kenya

Abebe Engidasaw, Addis Ababa City Council, Ethiopia

Catherine Goyder, Addis Ababa, Ethiopia

Miles Hardie, International Hospital Federation, London, UK

Trudy Harpham, London School of Hygiene and Tropical Medicine, UK

Arif Hasan, Orangi Pilot Project, Karachi, Pakistan

Harrington Jere, Human Settlements of Zambia, Lusaka, Zambia

Jung Han Park, School of Public Health, Taegu, South Korea

Marlene Kanawati, OXFAM, Cairo, Egypt

Hilda Kiwasila, Ministry of Lands, Tanzania

Tim Lusty, OXFAM, Oxford, UK

Karoline Mayer, Fundacion Missio, Santiago, Chile

Caroline Moser, Development and Planning Unit, London, UK

Sangwa Musinde, Project Sante Pour Tous, Kinshasa, Zaire

Kenneth Newell, Liverpool School of Tropical Medicine, UK

Lewnidah Ongari, Undugu Society, Nairobi, Kenya

Indumati Parikh, Streehitakarini, Bombay, India

Ishwarbhai Patel, Safai Vidyalaya Sanitation Institute, Gujarat, India

Trinidad de la Paz, Davao Medical School Foundation, Philippines

Trevor Peries, Public Health Department, Colombo, Sri Lanka

Gerry Pinto, UNICEF, Bombay, India

Brian Pratt, OXFAM, Oxford, UK

Olikoye Ransome-Kuti, Minister of Health, Nigeria

Rama Rau, Municipal Corporation of Hyderabad, India

Victoria Rialp, UNICEF, Brasilia, Brazil

Susan Rifkin, Liverpool School of Tropical Medicine, UK

Federico Rocuts, Departamento de Investigacion y Projectos, Bogota, Colombia

Alessandro Rossi-Espagnet, WHO Consultant on Urban PHC, Geneva, Switzerland

Pralom Sakuntanaga, Bangkok Metropolitan Administration, Thailand

David Satterthwaite, International Institute for Environment and Development, London, UK

Abdalla Sulliman, Islamic African Relief Agencies, Khartoum, Sudan

Iraj Tabibzadeh, WHO, Geneva, Switzerland

Jaime Galvez Tan, UNICEF, Manila, Philippines

Jember Teferra, Redd Barna, Addis Ababa, Ethiopia

Patrick Vaughan, London School of Hygiene and Tropical Medicine, UK

Miriam Vila, Ministry of Public Health, Havana, Cuba

Joe Wray, Colombia University Medicine Faculty, New York, USA

Khairuddin Yusof, Malaya University Medicine Faculty, Kuala Lumpur, Malaysia

Tables

Figures

1
Introduction

The slum: 400 baht a month for a family consisting of one
father, one mother, eight children, four dogs, ten cats, six
ducks, and ten million mosquitoes.

Morell and Morell, 1972

Redressing the balance

The urban poor live in the shadow of the city. Often the slums and shantytowns are physically dominated by the prestigious tower blocks of the city's

Fig. 1.1 A cellar offers shelter for a migrant family in Lima, Peru. Photography
courtesy of WHO.

central commercial areas. This 'official' city is the one known to business
people and tourists. The residents of the slums and shanty towns are also in
the shadow of the rest of the city in terms of quality of life and services, as
this book will demonstrate.

The International Conference on Primary Health Care held at Alma Ata
recommended that 'governments incorporate and strengthen primary health
care within the national development plans with special emphasis on rural
and urban development programmes' (WHO 1978). This recommendation is
reflected in the World Health Organization's (WHO) global strategy of
Health for All by the year 2000, which makes explicit reference to urbanization and its problems (WHO 1981). Since 1978, however, most of the initiatives, and especially the literature relating to primary health care (PHC),
have focused upon rural populations. One aim of this book is therefore to
redress this imbalance.

This imbalance arises partly from the fact that traditionally, cities have
benefited from a disproportionate share of resources, including those available for health care. The first inequity to be perceived by promoters of
primary health care was the obvious difference between the conditions of the
urban elite and the rural poor. Discussion of primary health care in cities of
the developing world should not be interpreted as advocating a shift of
attention from rural to urban areas, where the latter are already privileged.
The intention of this book is rather to emphasize the enormity of the urbanization problem, and to underline the profound inequalities among city
populations. Case studies are used throughout to illustrate the pertinent
issues. As we know more about some aspects than others, this is reflected in
the length of the chapters.

Some of the recent literature calling for initiatives in PHC for the urban
poor is restricted mainly to the demographic argument that more people in
cities means that more health resources should be allocated to meet their
needs. Considering that most health resources are already concentrated in
urban areas, this argument alone is not very convincing. The argument for
more PHC initiatives for the urban poor rests upon the *rapid rates* of urbanization and the *inequity* experienced within cities. Chapters 2, 3, and 4 focus
upon these issues while Chapters 5–14 deal with specific problems and
programmes.

Conceptually and operationally, the primary health care approach presently being implemented around the world is, with few exceptions, a rural
approach based on rural health information, rural social structures, and on
rural administrative frameworks. However, the case studies in this book
demonstrate a variety of possible initiatives in health care for poor urban
communities that will hopefully act as catalysts for future programmes.

Fig. 1.2 A street child of Ecuador. UNICEF estimates that there may be up to 40 million abandoned children in Latin America and the Caribbean. WHO photography by P. Almasy.

Why should organizations be interested in urban health development?

Recognition of a problem is one thing but the will to act is another. Tan (1985) addresses the question of why institutions and organizations should pay more attention to urban health development:

'The answers to this question are as varied as the types and persuasions of organizations currently involved in urban primary health care. For city government agencies, it is their duty and obligation to provide health care to the most needy sectors of the urban areas. For university-based health projects, the urban areas are the most accessible places for students' field studies and exposure. For convenience and necessity, the urban slum becomes the laboratory and experimental area to train students in community health. For civic organizations, religious groups and charitable agencies in the cities, what better way is there to *ease their conscience* and express their Christian works of mercy than through free clinics, deworming sessions, dole-out food rations, and relief goods. For social action and cause-oriented groups, the urban slums are where the poor, deprived, and oppressed are located and where efforts of conscientization and organization towards self-reliance find more meaning.

To date, many major urban infrastructure projects are being undertaken without including basic primary health care dimensions, which means that the projects do not necessarily improve the health conditions of the urban poor. The apparent scale of action has little hope of reaching all of the urban poor even by the year 2000 (WHO 1985). Increasing the population coverage of initiatives is discussed in Chapter 14.

Since 1983, WHO and UNICEF have been collaborating on problems and programmes in poor urban areas. UNICEF (1982) has reported on its urban basic service actions in 43 developing countries. In virtually all of these countries, various components of the primary health care strategy were supported. In many of them there was a very strong emphasis on primary health care itself, as for example in Peru, Ecuador, Colombia, Honduras, Brazil, Belize, Thailand, India, Indonesia, and Ethiopia. Workshops and conferences on urban primary health care have been sponsored by WHO and UNICEF, and these two organizations continue the 'conscientization' concerning the health conditions of the urban poor.

Turning to academic institutions, despite encouraging editorials in journals like *Social Science and Medicine* (Verhasselt 1985) and the *Journal of Tropical Paediatrics* (Ebrahim 1983) and special articles in the *Lancet* (Maxwell 1986), very few researchers are focusing their attention upon health conditions in poor urban environments of developing countries. This is reflected in the lack of good data on the subject. Verhasselt (1985) has stated 'there is a challenge here that must not be ignored'.

2

Urbanization and the growth of the urban poor

> In the next two decades, the world will undergo, as a result of the urbanization process, the most radical changes ever in social, economic and political life.
>
> Rome Declaration on Population and the Urban Future, International Conference on Population and the Urban Future, UNFPA, Rome, 1–4 September 1980

Indicators of urbanization

Urbanization in the sense of 'the relative increase in the urban population as a proportion of the total' (Teller 1981) and its consequences has been one of the most widely discussed social issues of the last three decades. At the beginning of the last century, the world's urban population totalled less than 50 million; today it exceeds 1.6 billion, and the United Nations (UN) predicts a figure of 3.1 billion by the year 2000. In developing countries it is estimated

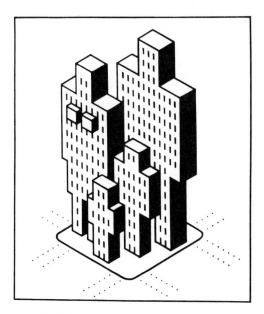

Fig. 2.1 More than half of the human race will be living in urban areas by the end of this century, says the UN Fund for Population Activities. *Source*: UNFPA.

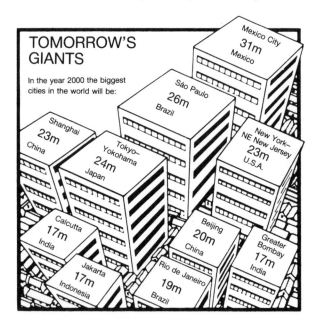

TOMORROW'S GIANTS

In the year 2000 the biggest cities in the world will be:

Mexico City 31m Mexico

São Paulo 26m Brazil

Shanghai 23m China

Tokyo–Yokohama 24m Japan

New York– NE New Jersey 23m U.S.A.

Calcutta 17m India

Beijing 20m China

Greater Bombay 17m India

Jakarta 17m Indonesia

Rio de Janeiro 19m Brazil

Fig. 2.2   By the year 2000 there will be 60 cities of over 5 million people, compared with 29 today. 61 per cent of urban growth will come from natural increase—the rest from migration. *Source*: Eaton, United Nations Graphics.

that 44 per cent of the population will be living in urban areas by the year 2000. Some Third World cities are expected to reach extremely large sizes by the end of the century: Mexico City, 31 million; Sao Paulo, 25.8 million; Rio de Janeiro, Bombay, Calcutta, and Jakarta each exceeding 16 million; Seoul, Cairo, Manila exceeding 12 million (World Bank 1984). Of the 26 cities of 5 million or more in the world in 1980, 16 of them were in developing countries. By the year 2000 the developing world is projected to triple its cities of this size to 45, out of a world total of 60 (see Table 2.1). However, only 23 per cent of the urban population in the Third World will be living in cities with 4 million or more inhabitants by 2000, i.e. most of the urban population will be living in smaller towns and cities (Hardoy and Satterthwaite 1986*a*).

Although the above figures are useful in indicating trends, certain caveats must be borne in mind when using them. The definition of 'urban' varies from country to country. For example, in Peru 'urban centres' are those with 100 or more occupied dwellings; in India they are settlements with 5000 or more inhabitants. Mostly, definitions fall between 1500–5000 people but there are undoubtedly problems when comparing urban populations. It is safer to examine the *number* of large cities for comparative purposes since

Table 2.1 The 60 urban agglomerations expected to have populations of over 5 million in the year 2000

| City (country) | Population (in millions) | | | City's population in 1980 as a percentage of: | |
	1950	1980	2000	National population	Total urban population
Africa					
Cairo (Egypt)	2.5	7.4	12.9	17.6	38.6
Addis Ababa (Ethiopia)	0.2	1.7	5.8	5.2	36.6
Nairobi (Kenya)	0.1	1.3	5.3	7.9	57.3
Jos (Nigeria)	0.4	2.7	7.7	3.5	17.0
Lagos (Nigeria)	—	1.2	5.0	1.6	7.9
Kinshasa (Zaire)	0.2	3.1	8.0	11.0	28.0
Latin America					
Buenos Aires (Argentina)	5.3	10.1	12.1	37.3	45.2
Belo Horizonte (Brazil)	0.4	3.0	6.5	2.4	3.6
Curitiba (Brazil)	0.1	2.1	5.2	1.7	2.6
Porto Alegre (Brazil)	0.4	2.5	5.0	2.0	3.0
Rio de Janeiro (Brazil)	2.9	10.7	19.0	8.5	13.0
Sao Paulo (Brazil)	2.5	13.5	25.8	10.7	16.5
Santiago (Chile)	1.3	3.9	5.6	35.1	43.6
Bogota (Colombia)	0.6	4.9	9.6	18.2	25.8
Guadalajara (Mexico)	0.4	2.8	6.2	4.0	5.9
Mexico City (Mexico)	3.0	15.0	31.0	21.4	32.2
Lima (Peru)	1.1	4.7	8.6	26.4	39.2
Caracas (Venezuela)	0.7	3.3	5.7	22.1	26.3
North America					
Chicago (USA)	4.9	8.2	9.3	3.7	5.1
Detroit (USA)	2.8	4.9	5.5	2.2	3.0
Los Angeles (USA)	4.0	11.6	13.9	5.2	7.1
New York (USA)	12.3	20.2	22.4	9.1	12.5
Philadelphia (USA)	2.9	4.9	5.5	2.2	3.0
East Asia					
Beijing (China)	2.2	11.4	20.9	1.2	4.7
Guangzhou (China)	1.5	3.4	5.7	0.4	1.4
Lanzhou (China)	0.3	2.7	5.5	0.3	1.1
Shanghai (China)	5.8	14.3	23.7	1.5	5.9
Shenyang (China)	2.2	3.4	5.3	0.4	1.4
Tianjin (China)	2.4	5.1	8.1	0.5	2.1
Wuhan (China)	1.1	3.2	5.0	0.3	1.3

Table 2.1   *continued*

City (country)	Population (in millions)			City's population in 1980 as a percentage of:	
	1950	1980	2000	National population	Total urban population
Hong Kong (Hong Kong)	1.7	4.4	5.9	91.0	100.0
Osaka-Kobe (Japan)	3.8	9.5	10.9	8.2	10.4
Tokyo-Yokohama (Japan)	6.7	20.0	23.7	17.2	21.9
Pusan (Rep of Korea)	1.0	3.1	5.4	8.2	14.9
Seoul (Rep of Korea)	1.1	8.4	13.7	22.1	40.6
Taipei (Taiwan)	0.6	3.3	6.8	0.3	1.3
South Asia					
Dacca (Bangladesh)	0.3	3.0	10.5	3.4	30.0
Ahmedabad (India)	0.9	2.5	5.1	0.4	1.6
Bombay (India)	3.0	8.4	16.8	1.2	5.4
Calcutta (India)	4.6	8.8	16.4	1.3	5.7
Delhi (India)	1.4	5.4	11.5	0.8	3.5
Hyderabad (India)	1.2	2.5	5.2	0.4	1.6
Madras (India)	1.4	5.4	12.7	0.8	3.5
Jakarta (Indonesia)	1.7	7.2	15.7	4.7	23.3
Surabaja (Indonesia)	0.6	2.4	5.4	1.6	7.8
Teheran (Iran)	1.1	5.4	11.1	14.2	28.4
Baghdad (Iraq)	0.6	5.1	11.0	39.0	54.6
Karachi (Pakistan)	1.1	5.0	11.6	6.1	21.4
Lahore (Pakistan)	0.9	2.9	6.6	3.5	12.6
Manila (Philippines)	1.6	5.5	11.4	10.8	29.9
Bangkok/Thonbori (Thailand)	1.4	4.7	10.6	9.9	68.6
Istanbul (Turkey)	1.0	5.2	10.8	11.5	24.1
Danang (Vietnam)	—	1.8	6.6	3.4	14.8
Europe					
Rhein–Ruhr area (Fed. Rep. of Germany)	6.9	9.2	8.6	15.1	17.9
Paris (France)	5.5	9.7	10.6	18.1	23.2
Milano (Italy)	3.6	6.7	7.9	11.8	17.0
Madrid (Spain)	1.7	4.6	6.0	12.3	16.5
London (United Kingdom)	10.4	10.0	9.2	17.9	19.8
Leningrad (USSR)	2.6	4.4	5.2	1.7	2.5
Moscow (USSR)	4.9	7.7	9.0	2.9	4.5

Source: Donohue (1982).

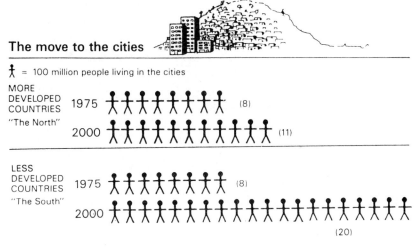

Fig. 2.3 The move to the cities. *Source*: WHO.

the same criterion is used for each nation. But even here the statistics for some countries are known to be inaccurate. Many countries have not had a census for years, so extrapolations are from very old data. For example, extrapolations from the last Nigerian census of 1963 showed Lagos to have a population of 1.2 million in 1980; UN revised figures suggest 2.8 million and recent national demographic household surveys point to a figure between 4 and 5 million (Hardoy and Satterthwaite, 1986b).

During any discussion of urbanization it is important to keep in mind that, contrary to what is often supposed, natural increase and not migration is the major factor for population increases. Natural increase is responsible for an average of 61 per cent of urban population growth in developing countries, compared to only 39 per cent from rural migration (UN 1980). Regardless of what policies are undertaken to affect internal migration, and their success or failure, governments have to come to terms with the increasing numbers of the urban poor who were born in the cities.

While the developing world is in general undergoing urbanization at a rapid rate, there are major differences between countries and regions. Donohue (1982) examines these differences in detail, but generally speaking, those regions with lower absolute levels of urbanization, such as Sub-Saharan Africa and parts of Asia, are in fact experiencing some of the highest levels of migration from rural to urban areas and the most rapid relative rates of urbanization. It is not simply the absolute level of urbanization (i.e. the per cent living in urban areas) that is important, but also the relative speed with which urbanization is occurring. Table 2.2 demonstrates

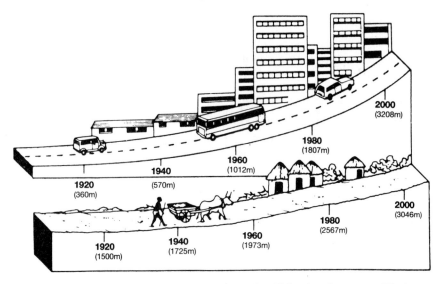

Fig. 2.4 In 15 years' time half the world's people will live in urban areas. The lower road shows the number of people in the countryside. The highway above adds urban dwellers to give the total world population. *Source*: United Nations Graphics.

Table 2.2 Urban population, total increase in urban population, and percentage increase, 1975–2000

Major areas of less-developed regions	Total urban population in 1975 (millions)	Total urban population increase 1975–2000 (millions)	% increase 1975–2000
Less-developed regions	838.4	1283.7	253
Northern Africa	38.2	71.6	287
Sub-Saharan Africa	66.0	185.9	382
Latin America	198.1	230.1	216
China	218.0	273.9	226
Other East Asia	30.6	34.0	211
Eastern South Asia	68.8	111.0	261
Middle South Asia	175.9	302.0	272
Western South Asia	21.2	42.8	302
Melanesia–Micronesia– Polynesia	1.1	3.4	409

Source: Donohue (1982).

the regional differences in the percentage increase in urban population from 1975 to 2000. This shows, for instance, that some of the most rapid changes are occurring in Asia, Sub-Saharan Africa, and the Pacific islands.

Traditional explanations of urbanization

In spite of a century and a half of debate, social scientists are still uncertain about the causes of urbanization. The two principal hypotheses which have been advanced are that rapid city growth and urbanization can be explained primarily by:

(i) unusually rapid rates of population growth pressing on limited land available for farming, pushing landless labour into cities; and
(ii) economic forces pulling migrants into the cities (Kelley and Williamson 1984).

These amounted to 'push' and 'pull' theories. For example, Engels (1845) thought that Manchester's booming growth in the early nineteenth century—and the urban decay associated with overcrowding—could be easily explained by the development of manufacturing under capitalism. Others thought that rural–urban migration, and thus town growth, was caused by land scarcity and enclosures. In short, Engels favoured 'pull', whereas others favoured 'push'. Moreover, recent events such as economic recession, drought, storms and floods, guerilla warfare, and nationalistic wars are contributing a further push to rural–urban migration in many developing countries.

Some more recent analyses of urbanization move beyond individual migrant motives and focus upon structural transformation in both rural and urban areas (see Moser and Satterthwaite 1985).

Growth of the urban poor

Moser and Satterthwaite (1985) provide a useful analysis of the characteristics and sociology of poor urban communities, and the rest of this chapter draws largely upon that work.

The rapid growth in cities has been accompanied by a rapid growth in the number of urban inhabitants who live in grossly substandard, overcrowded conditions without the funds for decent housing. The figures available from cities in developing regions indicate that in the 1960s and 1970s, slum and shanty town dwellers represented on average 30–60 per cent of the urban population. Estimates are that, at present, an average of 50 per cent of the urban population live at the level of extreme poverty, with this figure rising as high as 79 per cent in some cities (see Table 2.3). Assuming that in the year 2000 one-half of the urban population will still be of low income, over one billion people will be counted among the urban poor. 'These figures

Fig. 2.5 The urban poor often live in the shadow of the city. These shacks in Bombay are overlooked by the central business district of the 'official' city. Photograph by V. Virkud.

translated into human terms forecast very critical times ahead for most of the women and children living in the cities and towns of the developing world. Most of them will not benefit from the amenities, services, and economic opportunities that urban areas have to offer. They will be struggling for survival.' (Donohue 1982, p. 34).

Within the somewhat loose classification of 'low income people', there are some who have the least possibilities of acquiring an adequate income. Moser and Satterthwaite (1985) emphasize that disaggregation among low-income households shows that the poorest of the poor are the women-headed households. These households have the least access to income earning activities and yet have the most domestic childcare responsibilities. The fact that they have to manage the house and raise children as well as facing discrimination in the job market, greatly limits their possibilities of finding secure, reasonably paid employment. The growing number of *de facto* women-headed households in Third World cities is now increasingly recognized. Although there is a paucity of data on their extent, an estimated 30 per cent of all households worldwide are reckoned to be women-headed households with the proportion reaching more than 50 per cent in many Latin American cities (see Youssef and Hetler 1983).

It is important to understand why it is necessary to focus on urban poverty, despite the high concentration of all new productive investment in cities.

Table 2.3 Percentage of squatters and slum dwellers in selected cities (by region in descending order)

Region and city	Year	City population (thousands)	Slum dwellers, squatters (thousands)	% of slum dwellers, squatters to city population
Africa				
Addis Ababa	1981	1200	948	79
Casablanca	1971	1506	1054	70
Kinshasa	1969	1288	733	60
Nairobi	1970	535	177	33
Dakar	1969	500	150	30
Latin America				
Bogota	1969	2294	1376	60
Buenos Aires	1970	2972	1486	50
Mexico City	1966	3287	1500	46
Caracas	1974	2369	1000	42
Lima	1970	2877	1148	40
Rio de Janeiro	1970	4855	1456	30
Santiago	1964	2184	546	25
South Asia				
Calcutta	1971	8000	5328	67
Bombay	1971	6000	2475	41
Delhi	1970	3877	1400	36
Dhaka	1973	1700	300	35
Karachi	1971	3428	800	23
East Asia				
Manila	1972	4400	1540	35
Pusan	1969	1675	527	31
Seoul	1969	4600	1320	29
Jakarta	1972	4576	1190	26
Bangkok/Thonburi	1970	3041	600	20
Hong Kong	1969	3617	600	17

Source: Donohue (1982).

Governments frequently encourage this concentration through macro-economic pricing and sectoral policies. For instance, food prices are often subsidized for city consumers while rural producers are underpaid for their crops. Industrial development in cities is given preferential treatment to agriculture in terms of government incentives or credit, and the larger cities

Fig. 2.6   The poorest of the urban poor are those with least access to income-earning activities and most domestic childcare responsibilities—women-headed households. Photograph by V. Virkud.

receive most of the public investment in infrastructure and services. Macro-economic policies often serve to subsidize industrial production for domestic markets while penalizing exports. In economies which still have a large agricultural sector, this essentially taxes farmers and subsidizes urban invest-ment. The highly centralized structure of governments and the lack of power, funds, and resources in the hands of regional and local governments also helps concentrate resources and facilities in national capitals (see Hardoy and Satterthwaite 1986*a*).

However, although there is a high concentration of new productive investment in larger cities, with many government policies implicitly or explicitly promoting this, an examination of who actually benefits from such policies shows that it is the larger industrial and commercial interests and the upper income groups. The lower income majority receive little or no benefit from government investment in increased capacity for water supply, new telephone systems or new power stations since they do not receive these new services. The main beneficiaries of new highway systems in large cities are businesses and those with automobiles, not the low-income majority.

In addition, it is incorrect to make generalizations about urban resources

Fig. 2.7 Half of Bombay's 8 million people live in slum dwellings, many of them in Dharavi, reputedly the largest slum in Asia with over half a million people. Photograph by A. Charnock.

as smaller urban centres are often just as starved of new productive investment or public investment in infrastructure and services as rural areas. It is not so much the 'urban bias' which is the problem (see Lipton 1976), but large city bias; it is the middle- and upper-income inhabitants and larger enterprises in the larger cities within each nation which manage to ensure that government policies favour their needs (Moser and Satterthwaite 1985).

Causes of urban poverty

Just as there are different interpretations as to the causes of urbanization, there are different interpretations as to the causes of urban poverty.

Fig. 2.8   A Tale of Two Cities. *Source*: D. Brunner.

At its most simplistic, two very different stereotype interpretations have dominated development thinking as to the forms and patterns that social and economic development take in different Third World nations. The first is often known as the 'modernization' approach which views underdevelopment as a kind of backwardness. The conventional idea of 'modernization' is

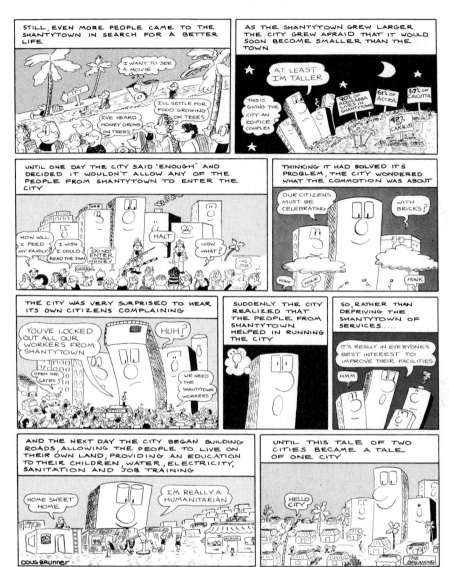

that Third World nations are undergoing a similar process of development to that experienced in the past by Western nations; it is such an idea which underlay Rostow's 'Stages of economic growth' which so influenced development thinking during the sixties and still retains some influence today. Behind the concept of this modernization process, there is the idea of

Third World nations having a dualist economic and social structure. The traditional (or backward) sector of this dual society contains the indigenous population, the rural communities, and the urban masses which are independent of the modern, advanced sector and 'outside the market economy and marginal to the national society and to the world as a whole' (Frank 1971, p. 41). The concept which has acquired widest usage to characterize this traditional or backward sector of the dual economy is the concept of 'marginality'. The problem for the traditional sector is seen as one primarily of adaptation to the modern system so that it can gradually be assimilated into the modern sector and the dual economy and social structure will cease to exist.

marginality

This concept of marginality has been used to describe particular characteristics of the urbanization process. The large numbers of migrants who are said to 'swarm' to the cities are termed 'socially marginal' while the fact that they are said to live in squatter settlements on the periphery of cities means they are also described as being 'spatially marginal'. As they may include a high proportion of illiterate and unskilled workers, the fact that they can find no secure employment is blamed on their own inadequacies; they are characterized as occupationally marginal and often said to contribute to 'over-urbanization'. Finally, the fact that they are thought to be outside any political organization or structure means they are frequently characterized as being 'politically marginal'. And the fact that there is frequently a physical separation between large squatter communities and city centres helped reinforce the assumption that their inhabitants are marginal to the city's economy, political structure, and society (Perlman 1976).

People's marginality came to be seen as a problem of adaptation to the system, identified in terms of the innate incapacity of the migrant himself or herself. In other words, the poor are seen as responsible for their own modernization, and, by implication, their own poverty (see Moser 1977, 1978, 1980).

historical materialist

By contrast, the structural (or 'historical materialist') interpretation argues that the problem of poverty is the consequence of a particular kind of economic development. The interrelation between the high rate of population growth, an increasingly capital intensive multinational development, and crisis in the agricultural sector has resulted in a substantial proportion of the labour force being 'marginalized' from stable and productive employment and thus from earning sufficient income to be able to afford to pay for adequate housing and basic services. The problem of people's poverty is then not one of characteristics and traits relating to the individual but is determined by the structure of the economy and the form its transformation has taken, determined both by national and international factors linked to internal issues of class and political institutions (see Moser 1977).

Perceptions of urban poverty

While this structural interpretation of development is now widely acknowledged at the macro-economic level, there is a curious tendency to cling to the modernization approach when analyzing the problems of poverty at the micro-level (Moser and Satterthwaite 1985). The latter authors point out that as recently as 1976, a UNESCO Expert Meeting on Urban Problems identified them in the following terms:

How many countries have had the melancholy experience of the rush to the towns at the moment when the government was preparing to launch a modest programme of urban development based on unrealistic growth rates extrapolated from inadequate samples . . . the ever increasing migratory movement—in practice beyond control—of families from rural areas attracted by the glitter and the fallacious promises of consumer society . . . These newcomers, potential parasites, consist in the main of large families—8 to 10 persons—only two or three of whom are economically active; and even these are, for the most part, without experience of the type of occupation, peculiar to the urban zone, in which they seek to work. The other members include, in addition to the aged and babies, women who can no longer take part in active life by working in farms or fields as they did before. Lastly, there are swarms of children, many of whom will not have access to education or, if they do will not complete the primary course . . .

This one way movement to the towns leads *a fortiori* to a very slow change. A country-dweller cannot adapt himself to town life overnight and squatters tend to settle on land that falls outside the scope of authorizations to build—areas regarded as unhealthy, subject to flooding, etc. In these conditions, the newcomer begins by reconstituting his former village background . . . Very soon, however, the physical or moral barriers of ownership and respect for neighbours are broken down by circumstances beyond the squatter's control. Promiscuity comes to be tolerated, and all that remained of the rules of community life rapidly vanishes. The squatter loses his peasant identity without on that account being accepted as a town-dweller. He becomes one of a mass of idlers where every man for himself is the rule. (UNESCO 1976, pp. 3–4.)

This UNESCO report perceives the urban poor as being socially, economically, politically and spatially marginal. Moser and Satterthwaite (1985) ask whether poor urban communities are really such a burden on the city and so incapable of improving the situation themselves, as the above report suggests.

Socially marginal?

The stereotype of the migrants attracted by the 'bright lights' has been seriously questioned by many detailed empirical studies both of migration processes and of poor urban communities. As noted earlier, not only are migration flows, by and large, rational responses by people to changing economic circumstances but, in addition, natural increase is often a more important factor than net in-migration to a city's population growth. It is not only the rural migrant who is found in slums or squatter settlements but also the urban born with long experience of living and working in urban centres. Indeed, in some squatter settlements, virtually all the inhabitants are relatively long-term urban inhabitants (see, for instance, Hamer 1981; Cuenya *et al*. 1984). This makes the view that the problem is one of the peasant unable to assimilate as a town-dweller appear increasingly untenable.

The description in the UNESCO document of the community in which a breakdown of moral order has occurred with promiscuity and deviant social behaviour evident is strongly reminiscent of Oscar Lewis's famous 'culture of poverty' concept. Lewis (1965) identified this influential conceptual interpretation of poverty in conditions of underdevelopment in terms of alienation rather than poverty itself and characterized those experiencing it in terms of some 70 inter-connected economic, social, and psychological traits relating to individual values, family and the nature of the family, and the organization of slum communities. The importance of the concept of a culture of poverty lies in the argument that it is a separate sub-culture from the national one which enables the poor to cope with a different environment, and that it has self-perpetuating mechanisms for its transmission to succeeding generations.

Economically and politically marginal?

Lewis failed to recognize the extent to which the poor do participate in the economic and political life of their city and to recognize the fact that their poverty was no more than the result of the lack of opportunities and resources rather than a consequence of a certain type of culture. Innumerable detailed studies of urban communities in different cities in Africa, Asia, and Latin America have demonstrated this, with authors like Peattie (1975) stating that 'the commerce of the central city streets and of the peripheral low-income *barrios* just as much as the commerce of big shops is part of the general economy of the city, not a separate sub-economy.'

The fact that such a high proportion of Third World cities' economically active populations are self-employed or work in small-scale enterprises which largely escape recognition, regulation, and government support led to the description of such activities as being the 'informal sector'. These include activities not only in producing goods but also in trade, commerce, and services. Despite the often low levels of productivity of many enterprises, and the critical constraints they face in terms of profitability and expansion, this is neither a marginal nor a disappearing sector, as was largely the case in nineteenth- and early twentieth-century Western Europe. In the cities of the Third World, this so-called 'informal sector' has been shown to be increasingly functional and useful to large-scale enterprises and to cities' economies

REACHING THE LIMITS

Around one billion people now live in Third World cities—a number that will double by the turn of the century and present an enormous challenge to city planners trying to cope with :

HOUSING

Cairo has 750 000 houses less than it needs—and the deficit is growing at 150 000 a year

HEALTH

Industry provides employment but also brings risks. 1 000 tons of pollutants fall daily on Greater Bombay

EMPLOYMENT

Most city people work— but don't earn much. In Bogota the top 5% of city dwellers get 30% of the income

TRANSPORT

Average traffic speed in Mexico is now half that in London or Paris.

Fig. 2.9 Reaching the limits. *Source*: United Nations Graphics.

in general in a number of ways. For instance, Portes (1978) has argued that the low cost of urban labour in most Third World cities (from which both national and multinational enterprises benefit) is underpinned by the fact that so many cheap goods and services are produced by this informal sector.

Spatially marginal?

Squatter settlements are often located on the periphery of cities because it is easier to escape forceful eviction from such sites. Illegal occupation of relatively low-quality and poorly located land sites—especially on publicly owned land or on land for which ownership is under dispute—is much safer than squatting on more valuable and better located land close to the city centre. Squatting on the latter would almost certainly lead to instant, forceful eviction.

The fact that such squatter settlements are spatially separate from the main city centre does not mean that their inhabitants are not closely integrated into the urban economy. Empirical studies have given many illustrations of how many residents in peripherally located settlements make long journeys to and from work, despite daily journeys of up to two hours or more each way. The level of organization within squatter settlements is one of the clearest indications of the sophisticated level of social organization in poor communities, which again goes against the conclusions of the UNESCO report. Many squatter communities are carefully planned prior to occupation; space is left for straight roads, paths, and community facilities. Space is also often left for social gatherings (see Turner 1976; Cuenya *et al.* 1984).

Fig. 2.10   The crowded conditions in many areas means that homes are formed in the most unlikely places. Photograph by P. Smith, OXFAM.

The urban poor as a promise and a threat

Since the 1960s, when squatter settlements were viewed as temporary eyesores and bulldozing was often the preferred policy response, many governments have come to realize that these settlements are functional to city economies and of benefit to city governments. The fact that they are largely self-organized, self-built, and self-managed reduces the responsibility of the state to provide adequate housing for the large labour pool within cities. And the goods and services that many provide in small informal enterprises serve the needs of capitalist enterprises and the consumer tastes of middle- and upper-income groups.

In practice, however, the urban poor are *still* sometimes perceived as a threat. Jumaoas (quoted in Tan 1985) suggests that the World Bank perceives the urban poor as both a promise and a threat. A promise because they constitute a vast reserve of cheap labour for export industries, and a threat because of 'upsurging urban poor mass movements which tend to disrupt international investments'. This kind of view was expressed by McNamara, the former Bank president, during an annual International Monetary Fund/World Bank meeting in September 1975:

> Historically, violence and civil upheaval are more common in the cities than in the countryside. Frustrations fostered among the urban poor are readily

exploited by political extremists. If cities do not begin to deal more construc-
tively with poverty, poverty may well begin to deal more destructively with
cities.

Summary

- Urbanization is occurring at rapid rates in almost all developing countries.
  African countries are now encountering the rapid rates of urbanization that
  Latin America experienced in the 1960s/1970s.
- Most urban growth is from natural increase within the cities with in-migration
  from rural areas playing a smaller role—although there are significant regional
  differences in this balance (e.g. natural increase is more important in Latin
  America while in-migration is the major cause of urbanization in Africa).
- This rapid urbanization is accompanied by rapid growth in the number of
  urban poor in shanty towns, squatter settlements, and slums. Many cities have
  about half of their population in such settlements.
- The 'modernization' concept of development which views underdevelopment
  as a kind of backwardness has characterized such poor urban populations as
  being marginal (socially, physically, economically, and politically). Using this
  interpretation the poor are seen as responsible for their own poverty.
- An alternative to the modernization approach is a structural interpretation. In
  this interpretation poverty is determined by the structure of the economy and
  influenced by national and international factors.
- There is often a failure to recognize the extent to which the poor actively
  participate in and contribute to the life of the city (particularly through the
  informal sector).

3

Characteristics of poor urban communities

Squatter settlements by definition and by city ordinance are illegal. Even the word squatter itself is vaguely obscene, as if somehow being penniless, landless and homeless were deliberate sins against the cannons of proper etiquette. But it is not squatters that are obscene. It is the economic circumstances that make squatter settlements necessary that are obscene.

Robert McNamara, World Bank President, during IMF–
World Bank meeting September 1975

Types of poor urban communities

The urban poor population is seldom homogeneous and the type of urban community affects the strategies and methods to be used in a community health programme. Is it an old slum, a resettlement area, a squatter settlement, or a mixed community of poor and middle-class families? Each of these would have its own characteristics (Tan 1985). In relation to the use of these different terms Rossi-Espagnet (1984) suggests the following definitions:

Shantytowns: once a commonly used term, but now considered pejorative, referring to the external view that the low-income settlements are only make-shift huts.

Slum: usually referring to the old, deteriorating tenements in the city centre (originating from the word slump meaning 'wet mire' where working-class housing was built during the British industrial revolution in order to be near the canal-based factories).

Squatter settlements: originally referring to the fact that the inhabitants squat on, or do not have legal tenure to, the land but now often referring to the new slums where the inhabitants sometimes do have legal title. 'Squatments' is contrived from squatter settlements to include a broader range of the new slums and not simply to imply that all the inhabitants in such settlements are squatting. Besides this familiar term, many adjectives have been officially applied to modify 'settlements', among them 'marginal', 'transitional', 'uncontrolled', 'spontaneous', 'sub-integrated', 'non-planned', 'provisional', 'unconventional', and 'autonomous'.

A general typology of low-income settlements is provided in Table 3.1.
There are also many different local terms for low-income urban settlements:

Brazil: *favelas, alagados, vilas de malocas, corticos, mocambos*
Peru: *barriadas, pueblos jovenes, barrios marginales*

Table 3.1 General typology of low-income settlements

Type	Land acquisition	Tenure	Land and physical characteristics/ propensity for upgrading
(1) Irregular settlements (1a) Squatters	invasion of public or private land	*de facto* or *de jure* 'ownership'	Periphery where security of tenure then up-grading is likely. Otherwise static shantytown structures with little consolidation.
(1b) Illegal sub-divisions	sale of private land sale of customary land	owners, although titles may be imperfect	Periphery. In various phases of consolidation and upgrading through 'self help'.
(2a) Shanty towns	squatter or, more usually, renters	rental	Mostly down town and around city centre/small plots. Structures with few public amenities and little prospect of their provision. Little likelihood of self-help improvements because: (a) no security of tenure; (b) difficulty of creating an investment cash surplus for improvement (because of rental outgoings); (c) very small plot sizes.

Table 3.1 *continued*

Type	Land acquisition	Tenure	Land and physical characteristics/ propensity for upgrading
(2b) 'Street sleepers'	may have regular sleeping places	none	Down town/inner city. Minimal shelter, removed daily. Often associated with workplace.
(3) Tenements	converted large houses or purpose-built tenements	rental	Mostly down town. An increasing proportion of new rental accommodation is located in older irregular settlements (1a, 1b). Single room per family and share services.
(4) Public housing projects			
(4a) Complete units	government purchase and sales	owner or rental	High density, good amenity, but relatively expensive.
(4b) Incomplete units —site and services	government purchase and sale	owner	Periphery. Basic services mostly installed but house construction varies,
—'core' units	(sometimes interrelated agencies sponsorship)	owner	Self help and mutual aid to improve external house structure.

Source: WHO Environmental Health Division, unpublished data.

Venezuela: *barrios, ranchos*
Mexico: *colonias proletarias, colonias para-caidistas, jacales, ciudades perdidas*
Chile: *poblaciones, callampos* ('mushrooms')
Argentina: *villas miseria*
Colombia: *barrios clandestinos, tugurios, invasion*
Morocco: *bidonvilles*
Tunisia: *gourbivilles*

Turkey: *gecekondu*
India and
Pakistan: *bustees, jompris, chawls, ahatas, cheris, katras, juggies*
Zimbabwe: *periurban septic fringes*

Although it is difficult to generalize about the characteristics of poor urban environments, certain indicators in addition to income levels are particularly useful. The following selected characteristics fall into two categories: (a) demographic and socio-economic, and (b) physical and institutional.

Demographic and socio-economic characteristics

The most vulnerable populations are often found where population density is highest. For example, population density in slum areas in Calcutta and Manila is four or five times higher than the averages for the entire city (Austin 1980).

Another helpful indicator is household density, defined by either square metres of floor space or the number of rooms per household member. For example, Bangkok slums average about 3.5 people per room, far above the city average (Morell and Morell 1972).

In addition to examining density figures, population growth rates should be examined, as disproportionately accelerating population growth usually implies a pressure upon services. For example, although Manila City grew at a rate of 1.5 per cent per annum between 1960 and 1970, the slum areas of Manduluyong, Caloocan City, and Quezon City expanded at a rate of 7 per cent (World Bank 1975). However, Austin (1980) warns of the pitfalls that come from looking at a single indicator. For example, Cali, Colombia, and Calcutta slums show below average growth rates. An examination of the population density statistics, however, reveals rates four to six times greater than city averages. The low growth rates probably indicate a more mature or stabilized poverty zone full to physical capacity. Other, new poverty areas are probably expanding in these circumstances.

Poor urban communities are often characterized by in-migration from rural areas and other urban centres. For example, in Sao Paulo it was estimated that, of the 3.2 per cent per annum population growth between 1920 and 1940, one-half could be attributed to migration. In the following decades the growth was 6.3 per cent per annum, of which two-thirds was caused by migration (Escola Paulista 1975). However, as discussed in the previous chapter the importance of natural increase from the population already in the city must not be underestimated.

The relatively high proportion of young people in poor urban families exacerbates the dependency ratio. A study of the low-income areas of Manila disclosed that 64 per cent of the population was under 20 years of age,

Fig. 3.1   The urban canyons of Bombay. Far from being nests of crime and revolution, most squatter settlements consist of ordinary people, able and willing to play their part in urban development. Earthscan photograph by M. Edwards.

whereas in the Manila metropolitan areas only 47 per cent were in that age group (Basta 1977). In the Bangkok slums the group under 14 years of age made up 45 per cent of the slum population (Morell and Morell 1972).

   Mothers in employment is another characteristic of poor urban communities. For example, although in other areas the percentage is much lower, more than half of the mothers in low-income families studied in Bangkok slums were employed outside the home, and the proportion is similar in Rio de Janeiro (Getulio Vargas Foundation 1975).

Physical and institutional environment

Slums and squatter settlements are frequently beyond the reach of public water services. For example, only 28 per cent of the inhabitants of the Tondo Foreshore, Manila and 38 per cent of the low-income group in Sao Paulo had access to piped water (Austin 1980). Water that is available is expensive. In Cali, Colombia, the price of water purchased from a vendor was about ten times greater than the price of an equal volume of tap water. In the Klong Toey slum in Bangkok the cost per month of water was equivalent to four days' wages (Morell and Morell 1972).

Low sewerage connection rates often also characterize poor urban environments. Drainage and runoff from storms, periodic floods, and soil erosion commonly exacerbate the sewage disposal problems in slums (see Ch. 9). In the poorest section of Calcutta there was no sewerage at all, whereas 54 per cent of the Calcutta Corporation has adequate sewers (Austin 1980). In Manila there are 29 sewer connections for each 1000 people, although in the Tondo district the connection rate is only 15 per 1000 (World Bank 1975). In Sao Paulo, 57 per cent of households studied were connected to sewers whereas 41 per cent used ditches (Escola Paulista 1975).

Poor housing is the most visible sign of a low-income area. The lowest quality housing is generally found in squatter communities in which the threat of eviction reduces the incentive to invest in home improvements.

In addition to the above features there is a general lack of services, such as refuse disposal and electricity. As the following chapters indicate, the

Fig. 3.2 Homes of new migrants from the countryside straggle up the hillside above Lima. WHO photograph by J. Bland.

inequality and inaccessibility of health-care services is a basic problem in these areas.

The following descriptions of poor urban communities highlight some of the above characteristics.

Examples from selected cities

Addis Ababa (Ethiopia)

With a current population of about 1.5 million, Addis is the largest city in East Africa. A survey carried out by the World Bank in 1978 classified 79 per cent of the city population as living in low-grade, congested settlements of one-storey buildings, under conditions unfit for human habitation (Hailu 1978). Houses are made of staves of wood covered with mud, straw, and sometimes cow dung. The heavy rains gradually wash away the mud plaster which needs continual maintenance. Often, low-income areas can only be approached by narrow rugged pathways which are inaccessible for service vehicles. Facilities such as sanitation, piped water, and refuse collection are lacking. At present 30 per cent of urban dwellers have no water supply or access to a common standpipe, only 35 per cent of the solid waste generated

Fig. 3.3   79 per cent of Addis Ababa's population live in one-storey buildings under conditions unfit for human habitation, according to a World Bank survey. Photograph by C. Goyder.

in the city is collected and disposed of satisfactorily, and in 1977 24 per cent of all households had no toilet facility at all (Goyder 1985).

Bangkok (Thailand)

Bangkok has a population of 5 million people, it dominates the urban structure of Thailand by absorbing 70 per cent of the total urban population. The growth in the population is mainly due to a continuous rural–urban migration. Migrants gravitate to 410 slum areas scattered in the city, and make up approximately 25 per cent of the city population. The general health of these communities is poor in spite of the large health resources the city has to offer. Many of the slums and peri-urban slum dwellers have little access to essential health-care services (Sakuntanaga 1985). Many of the unauthorized settlements are on swampy land prone to flooding and along railway lines or canals. It is common for households in these settlements to receive the landowners' permission to build a makeshift house but only temporary tenure is given, so occupants can be removed when the owner wants to sell or develop the land. A major eviction programme in Bangkok threatened people living in low-income settlements. By 1981, 61 600 had been evicted and 148 600 were in the process of being evicted (Nitaya and Ocharoen 1980; Boonyabancha 1983; Wegelin and Chanond 1983).

Bombay (India)

Bombay as a port, industrial centre, and capital of the Maharashtra state is a mini-India with inhabitants from all cultural and ethnic groups. Of its 8 million inhabitants, 2.8 million live in slums or 'hutments', without access to basic amenities. Of these, the Dharavi slum, with over half a million people (1981 census), is considered to be the largest slum in Asia. The living conditions in the slums are bad and in some cases defy description (Rossi-Espagnet 1985). The slum dwellers live in appalling conditions with poor ventilation, poor drainage, overcrowding, and lack of basic amenities which are reflected by the effects of unhygienic conditions among the women and children. Many of the women complained of generalized weakness and anaemia (70–75 per cent), 50–60 per cent suffer from chronic malnutrition, avitaminosis, recurrent gastroenteritis and helminthic infections. The infant mortality is considered high in Bombay, the main causes being pneumonia and other communicable diseases. Malnutrition, blindness, and paralysis are common causes of morbidity. There are also other equally important social conditions which affect the population, such as sexually transmitted diseases, early teenage pregnancy, smoking and drug addiction among school children.

Many of the slum dwellers work outside their place of residence, receiving poor wages in white- and blue-collar jobs. Many jobs are unprotected. Others are employed in the 'informal' sector. Their level of education, on the whole, is poor (Sukthankar 1983).

Fig. 3.4   Pavement dwellers in Bombay sometimes build fairly permanent shelters. Photograph by V. Virkud.

Guayaquil (Ecuador)

Guayaquil is the largest city of Ecuador, with a current population of 1.3 million expanding from 26 000 inhabitants in 1950 and expected to rise to 3.1 million in the year 2000. On the edge of the city's commercial centre there are the inner-city slums, the *tugurios*. Here the houses, either subdivisions of decaying middle-class housing or purpose-built tenements, accommodate up to 15 people in appalling, overcrowded, and insanitary conditions. But much of the city's rapid growth in population since 1930 has been housed in squatter communities built over an area of tidal swampland to the west and the south of the city centre which has little commercial value in its natural state. Population in this area, known as the *suburbios*, has grown far more rapidly than the rest of the city. In 1975 it was estimated to contain 60 per cent of the city's population. The major part of the *suburbios* are kilometre upon kilometre of small, bamboo and timber houses standing on poles above mud and polluted water. The houses are connected by a complex system of timber walkways which also link to the nearest solid land. In some areas, land is a 40-minute walk away (Moser 1982). In certain areas, the sewage disposal system, poses insurmountable technical problems; the drinking water supply system is most inappropriate, its distribution work is by tank at a cost 20 times higher than if it were piped (Rossi-Espagnet 1985).

Jakarta (Indonesia)

In Jakarta, with a population of 6.5 million, the poor-income group can be difficult to identify, but certain indicators based on income and rice consumption have proved to be useful and indicate that 50 per cent of the population are poor. Many of them are unemployed or work as servants, unskilled labourers, small traders, refuse collectors, and poorly paid civil servants. Many make their permanent homes in urban slums, while the seasonal workers move around according to their jobs. No definite information is available on the type of diseases which are prevalent among the poor, but communicable diseases and malnutrition are common. The overall health status of the city population is poor with high infant mortality, crude death and fertility rates. The influence of the health environment is reflected in the short life expectancy of the people, 50 years. (Sudharto 1983).

Kuala Lumpur (Malaysia)

A 1978 census revealed an estimated 250 000 squatters comprising 25 per cent of the population of Kuala Lumpur. They are distributed over 140 urban *kampungs*. It is estimated that their population has now increased to 350 000 (Yusof 1985*a*). The squatters are mainly distributed according to ethnic groups, with Chinese and Malay predominating. Communal standpipes are available for 65.6 per cent, and 56 per cent use pit latrines, 22 per cent bucket latrines. Many of the houses are made of unpainted wood with cement floors and are generally satisfactory, with less than 3 per cent being dilapidated. The situation of the squatter families in Kuala Lumpur appears to be far better than those of other cities previously described, with 97 per cent of heads of households being employed.

Manila (Philippines)

A 1978 report suggested that there were a total of 328 000 squatter families with a total population of close to 2 million inhabitants living on 415 sites dotted through the whole urban region. These do not include families living in legal but otherwise substandard housing. Manila's squatter settlements vary in size from the very large Tondo Foreshore area with 27 600 families, or Bagong Barrio in Caloocan (part of Metropolitan Manila), with 16 800 families, to mini squatter settlements. Estimates suggest that only around 12 per cent of Manila's population can afford to buy or rent a legal house or flat on the open market. The greatest single determinant of housing costs is land costs. With the low incomes of most Manila households these two factors are at the heart of the housing problems and the extensive squatting. Yet land in Manila is very expensive not because it is scarce but because it is hoarded. An aerial view shows large tracts of unused land surrounding the city and even within the urban core. The Task Force on Human Settlements estimates that

64 per cent of the wider metropolitan area was still open space in 1973 (Keyes 1980).

Mexico City (Mexico)

One of the largest cities in the world, Mexico City has increased from the 1.2 million of 1930 to the present population of 17 million inhabitants and is expected to rise to 32 million by the year 2000. There is an average of more than 1000 immigrants per day coming to the city from all parts of rural Mexico or from other cities in search of work and opportunities. The squatters are called 'parachutists' as they mostly arrive overnight as if fallen from the skies—their living conditions are poor or very poor, overcrowding is common and at times extreme, and the infant mortality rate in the poor areas can be up to three times higher than in the rest of the city. Protein–energy malnutrition is an important contributory cause of death among children between the ages of one to four years (Fajardo 1983). At least 7 million people live in some form of uncontrolled or unauthorized settlement. Until the 1940s most lower income groups lived in rented rooms in custom-built slums (*vecindades*) or in subdivided middle- and upper-income housing. Since the 1940s, most new low-cost housing has been in unauthorized developments which usually developed in one of two ways. Either landowners or real estate companies sold illegal subdivisions, or land allocated to rural communities under the long-standing agrarian reform was illegally subdivided and sold. In both circumstances the occupants' tenure is insecure since the subdivisions contravene official regulations and the infrastructure and service standards demanded by such regulations are not met. The land on which they are developed is frequently ill-suited to residential development because it is rocky and hilly, with unstable subsoil or dry and dusty environments (Connolly 1982).

Nairobi (Kenya)

The estimated population of Nairobi in 1979 was 827 800. An estimated 110 000 unauthorized houses accommodate around 40 per cent of the city's population (Amis 1982). These include unauthorized shelters built on land which is illegally occupied (which could be designated squatter) and unauthorized shelters on land either owned by the developer or used with the permission of the landowner. Mathare Valley, one of Nairobi's squatter areas, grew from 4000 inhabitants in 1964 to 200 000 in 1985 (Ongari and Schroeder 1985), and other areas with illegal developments grew at comparable rates. The unplanned and unauthorized settlements are typified by 'houses' of simple and usually low-quality construction, often in a poor state of repair, with inadequate or non-existent piped water or sewage disposal facilities. In addition, the settlements have no access roads and no public lighting, and population densities are very high (Njau 1982). 'Slum upgrading' is under way in Mathare. 'Nice looking stone houses are built whereby the rent is too high to afford for the present slum dwellers. So they are forced to move into another slum area which is further away from the city centre and the industrial area, forcing the people to spend more on bus fares or to walk longer distances when looking for casual labour.' (Ongari and Schroeder 1985.)

Nouakchott (Mauritania)

A small town with 5800 inhabitants in 1965; it had 135 000 in 1977, and estimates suggest that now it has 250 000 people or more. Most of this incredibly rapid growth in population has been housed in illegal shanty/tent settlements. The government has distributed 7000 plots of land since 1972 but these have received little or no services. An estimated 64 per cent of the population now live in largely self-built communities on these plots and in illegal settlements. More than two-thirds of the city's inhabitants have no direct access to water. Frequently, water has to be bought from a water merchant, with no guarantee as to quality and with the price up to 100 times that paid by those with piped water connections (Theunyck and Dia 1981).

The mixed nature of poor urban communities

It is important to recognize differences *between* and *within* poor urban communities. Taking the physical neighbourhood as a real or potential social entity sometimes disregards the structural realities of a highly compartmentalized society. There are difficulties in implementing any programme in a socially heterogeneous community. In addition to differences based on education, occupation, and income, there may well be differentiation on the basis of language, religion and caste. This can interfere with an integrated approach as different groups within a neighbourhood may find it difficult to co-operate, or may end up competing for control over assisted activities and benefits (Singh 1980). Bosnjak (1985) has described how, as in rural areas, the dependency of poor groups on the stronger ones may be an obstacle for the channelling of programme benefits to the most vulnerable groups. She suggests that to overcome this obstacle in its initial stage, the programme has to offer goods and services which are scarce among the groups with the greatest needs and which are less attractive to better-off groups.

In terms of differences between communities, Tan (1985) discusses the characteristics of the different types of poor urban communities in the Philippines. Old slums, like Tondo of Manila, Agdao of Davao, or Pasil of Cebu have the characteristics of rural migrants before the martial law period. The new slums are those communities to which the majority of residents migrated after 1972. Examples of the latter communities are the new slums of Cotabato and Zamboanga where thousands of Muslim refugees moved in at the height of the Muslim war of liberation in 1974–6, and the slums of Butwan and Davao which receive displaced farmers and the victims of militarization.

The power structure within a slum should always be borne in mind. 'In slums where there is a mixture of middle class and the poor, the feudal relationship between these two classes often persists. The middle class, wittingly or unwittingly, dominates over the poor majority in many aspects of everyday life. Thus when basic services are made available, it is the middle class who mostly benefit from them.' (Tan 1985.)

Yusof (1985*b*) similarly points out the importance of understanding the differences between the squatter and slum dwellers of a city. In the Malaysian context, he points out that the urban squatters are illegal occupants on land which belongs to the government or is under private ownership. Some may be given temporary occupancy licences. Usually they originate from economically depressed areas of the city. Unable to find shelter as legitimate tenants or home owners near their place of work, they build temporary shelter in which they can enjoy security of tenure within the city or the periphery of the town. Common sites for squatter settlements are along railway lines and river banks. In time, they tend to assert a moral claim to the site. What has created complications and generated animosity against the squatters is the emergence of squatter landlords and affluent squatter groups. These people create a complex pressure group on the squatters, renting rooms, providing jobs and, in time, even achieving control of the community through political power. They are articulate and forceful in putting forward their rights, and often have links with political groups. Their twin objectives are for personal profits and political advancement. A recent survey of squatters in Kuala Lumpur showed that 4.4 per cent of squatters are in social class I, i.e. professional groups such as architects, engineers, doctors, and senior civil servants. It is important to understand such differentiation because, in some squatter settlements, the health status of the community may be as good as the rest of the city dwellers. However, in general the environmental health of the community is poor (Yusof 1982).

Similar difficulties in identifying the poorest of the poor were experienced in Kinshasa, Zaire. Musinde and Lamboray (1985) describe how the project 'Santé pour Tous' began by implementing Health Zones in what appeared to be the poorest area at the periphery of Kinshasa which had no health services. However, they gradually discovered that the peripheral communities were not the most vulnerable. This was because:

- many women can leave these areas to look after their fields which are located close by,—this makes an important contribution to the family diet and income;
- those living in the peripheral areas are mostly small property owners; tenants are to be found in the city centre;

- the environment is healthier, no blocked sewers, no stagnant water; however the lack of drinking water was a problem;
- the number of unmarried mothers and children cared for by other family members is relatively low;
- transport to the city centre, which is 20 km away, is difficult so that only the most able-bodied live in the peripheral areas.

Musinde and Lamboray concluded that the decision, which seemed obvious at the outset, to provide services in peripheral areas was perhaps not the wisest, and that the most vulnerable communities are, in fact, to be found in the less salubrious old quarters in the centre of Kinshasa. In these quarters the problem becomes more complex because of the greater concentration of polyclinics and private dispensaries, combined with the greater diversity in socio-economic standards of residents living in the same part of the town centre.

Identifying the poor

One of the best ways of discovering the characteristics of a poor urban community is to ask the people themselves. This kind of 'participatory diagnosis' leads to an understanding of poor people's felt needs and of particular conditions of the most vulnerable groups which are usually left out in programme designs based on statistical averages (see Table 3.2). Both general 'life conditions' diagnosis and 'specific sectoral' diagnosis (with a focus on,

Table 3.2 Community diagnosis of the vulnerable groups in Coatzacoalcos, Mexico

Vulnerable groups:

- '*Playeros*' beach people who are not house owners
- Unemployed or occasionally employed people
- Families with undernourished children
- People with bad housing
- Single mothers

Common problems:

- Lack of drainage
- Lack of piped drinking water
- Presence of garbage and sewage
- Children's bad health
- Lack of pre-school services
- Working children and school drop-outs.

Source: Bosnjak and Cayon (1985).

Fig. 3.5   One characteristic of many poor urban communities is surprising hope and joy. Photograph by Pearson, OXFAM.

say, drainage or nutrition) can be performed successfully by the community. Bosnjak (1985) describes how communities in low-income areas of Coatza-coalcos, Mexico, have participated in the collection of information for the definition of priority objectives and in the design and implementation of projects, e.g. the selection of their health workers and the negotiation of solutions to their problems (land ownership, drainage, education, etc.). This type of approach emphasizes the fact that the community members are, as Bosnjak calls them, 'experts in their own reality'. Examples throughout this book demonstrate that, particularly in the complex urban environment, it is programmes using such participatory diagnosis that prove to be most successful.

Summary

- Low-income settlements include squatters, shanty towns, slums (tenements), public housing projects, and street dwellers.
- Useful demographic and socio-economic indicators of low-income urban set-tlements include high population density, high household density, disproportionately accelerating population growth, significant in-migration, high proportion of young people, and many mothers in employment.

- Typical characteristics of the physical and institutional environment include poor access to water and sanitation facilities, poor housing, and a general lack of other services.
- However, it is important to recognize the heterogeneity of poor urban communities and the implications this has for health programmes. Differences between and within communities are important.
- The structural realities of a highly compartmentalized society must not be disregarded. The power structure within communities often leads to difficulties in identifying and reaching the 'poorest of the poor'.
- Participatory diagnosis can help overcome such obstacles and enables a programme to reach groups with the greatest needs.

Fig. 3.6 Illegal activity. *Source*: D. Brunner.

4

Health problems of the urban poor

The urban poor are at the interface between underdevelopment and industrialization and their disease patterns reflect the problems of both. From the first they carry a heavy burden of infectious diseases and malnutrition, while from the second they suffer the typical spectrum of chronic and social diseases.

Rossi-Espagnet, 1984

The wide range of diseases of the urban poor

Good descriptions of the common health problems of the urban poor in the developing world can be found in Ebrahim (1984), Rossi-Espagnet (1984), and Harpham (1986). There are three groups of factors which are detrimental to health which operate heavily against the urban poor. The first includes direct problems of poverty such as low income, limited education, and insufficient diet. The second relates to man-made conditions of the urban environment, such as overcrowding, poor housing, industrialization, pollution, traffic, and a general increased exposure to infectious diseases. The third is the result of social and psychological instability and insecurity. Figure 4.1 attempts to summarize these health problems of the urban poor.

The excessive vulnerability of the urban poor and their exposure to pathogenic agents means that infectious diseases and malnutrition are severe health problems in slums and shanty towns. Often an imported reservoir of infection is continually replenished by rural–urban migration, which in turn reinforces local transmission, for example, as in the case of malaria.

Urban malaria is still a significant problem (see Yang 1982a, b). The breeding of *Anopheles* mosquitoes, the carrier of malaria, is made possible by the stagnating water collections from rain and other sources which are common in poor urban areas as a result of water mismanagement, lack of drainage and sanitation, and general environmental neglect. Epidemics of other vector-borne diseases, such as dengue haemorrhagic fever and increased transmission of filariasis, have occurred in poor urban areas. This is associated with the need of households to store water in iron drums or large earthenware containers leading to the breeding of the mosquito *Aedes aegypti* (vector of dengue and yellow fever), and the accumulation of waste water around settlements that favours the mosquito *Culex pipiens fatigans*, an increasingly important vector in the transmission of filariasis.

Tuberculosis is highly prevalent in the slums and shanty towns of the developing world, where infection is frequent and early. Malnutrition often

THE DREAM: HEALTH IN THE CITY

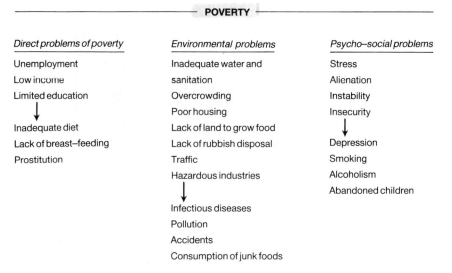

──── POVERTY ────		
Direct problems of poverty	*Environmental problems*	*Psycho–social problems*
Unemployment	Inadequate water and	Stress
Low income	sanitation	Alienation
Limited education	Overcrowding	Instability
↓	Poor housing	Insecurity
Inadequate diet	Lack of land to grow food	↓
Lack of breast–feeding	Lack of rubbish disposal	Depression
Prostitution	Traffic	Smoking
	Hazardous industries	Alcoholism
	↓	Abandoned children
	Infectious diseases	
	Pollution	
	Accidents	
	Consumption of junk foods	

Fig. 4.1 For many, poverty forms a barrier to the dream of a healthy life in the city.

occurs both as a cause and as an effect of diarrhoeal diseases (see Ch. 6). The scarcity and contamination of water supplies and the lack of sanitation and appropriate sewage disposal make diarrhoeal diseases one of the most important health problems in poor urban areas (see Ch. 9). In one year alone (1980) over 2 million urban children under five years of age died as a consequence of diarrhoeal dehydration (*Assignment Children* 1983). Disease incidence and mortality are also influenced by factors such as overcrowding, poor housing conditions, density of insects and vermin, lack of rubbish disposal, poor personal hygiene, contamination of food, low literacy, and inappropriate weaning and other feeding practices. A variety of intestinal parasites is usually present, with *Ascaris* and *Trichuris* often observed at higher levels than in corresponding rural populations (Richardson *et al.* 1968; Benyoussef *et al.* 1973).

In such crowded settlements there is also the danger of meningococcal meningitis and in many of the squatter areas of Latin America the incidence of leishmaniasis has been rising (Ebrahim 1984). A high incidence of preventable infections in children, such as measles, whooping cough, and polio suggests that adequate coverage with immunization is a priority in such overcrowded communities (Ebrahim 1984).

The man-made conditions of the urban environment cause particular health problems for the urban poor. Environmental pollution, which is a

Fig. 4.2   Pollution affects the poor. *Source*: OXFAM, Canada.

widespread problem for all urban people, affects the poorest more severely, since most of them live at the periphery where manufacturing, processing, and distilling plants are often built, and where environmental protection is frequently weakest (Rossi-Espagnet 1984). In 1984 the escape of lethal gas in Bhopal, India, which led to the death of 2500 people, and a gas explosion in Mexico City demonstrated the vulnerability of squatters who 'live next door to disaster' (Davis 1984; Centre for Science and Environment, New Delhi 1985). Herbert and Teas (1985) emphasize that urban, Third World children who are malnourished are especially likely to suffer ill-effects from toxic materials exposure.

Social and psychological problems as a result of political, economic, and social instability form another group of health problems for the urban poor. The protective structure of the local communities and extended family is generally replaced by the nuclear or single parent family unit in the city. Single parent households, often headed by a woman, are common, and with the need for such women to work the children are almost inevitably neglected.

The children may have to contribute to the family income, working under precarious conditions, and are exposed to accidents and abuse. UNICEF estimates that there may be up to 40 million abandoned children in Latin America and the Caribbean alone (Tacon 1981). The existence of so many children has been called 'a growing urban tragedy' by Agnelli (1986) who discusses street education as a response to the problem.

In contrast to the relatively stable and homogeneous rural village, the migrant to the city finds a society that is very heterogeneous, culturally and linguistically; transient and mobile; opportunistic and restless. Prostitution (often involving children), venereal diseases, drug addiction, tobacco smoking, and alcoholism are growing problems in poor urban areas (Rossi-Espagnet 1984). Zwingmann (1978) has reviewed the mental effects of the changes experienced on moving to the city, and Egdell (1983) examines a poor urban area of Cali and a densely populated area of Manila in an analysis of mental health care in the developing world. Alcoholism and depression are often found to be the most frequent and severe mental symptoms in these poor urban areas (WHO 1984a).

Care of the handicapped is rarely mentioned in the literature on poor urban communities. Pascual (1984) presents an unusual study on the disabled in a deprived urban community. In Bacolod City, Philippines, a programme was established in 1981 which maximizes the community's and family's self-reliant roles in relation to the physical, emotional, mental, and social needs of the community's disabled. UNICEF (1985) reviewed a number of other programmes for disabled children in poor urban environments.

The presence of large hospitals and out-patient departments has almost certainly had a depressing effect upon the development of a more coherent health services infrastructure in cities. Networks of health centres and health posts such as those found in rural areas rarely exist in urban situations. In practice, many people in the city go to pharmacists or drug sellers for basic advice and only visit hospitals when there is an emergency. This inevitably leads to a highly curative approach to health care. In addition, traditional practitioners and practices are commonly used among the poor in many cities (Odejide *et al.* 1977; Good and Kimani 1980; Heggenhougen 1980; Gelfand *et al.* 1981; Mahaniah 1981; Davidson 1983; Hirst 1983). Imperato (1979) has suggested that folk medicine has survived in modern urban centres not only because large numbers of people have faith in it, but also because as a system of medical care it has not remained rigid. It has adapted itself to the new urban scene.

Studies which have examined the health of the urban poor may be categorized into those which examine health only in the shanty or slum and make no comparisons with other areas, compare the health of the urban elite with that of the urban poor (i.e. intra-urban differences), and make rural–urban comparisons.

There is a variety of studies in each category covering a range of cities and countries. Each category is now examined in turn.

Studies of health in slums and shanty towns

Notable among studies which focus upon the slum or shanty alone and give an overall picture of conditions and health problems, are the surveys by the Morells (1972) in Bangkok, Burton (1976) in Lima, Bennagen (1981) in Manila, the Marga Institute (1982) in Colombo, Kothari *et al.* (1983) in Bombay, and Tekce and Shorter (1984) in Amman. Other studies focus upon certain aspects of health. For example, Datta Banik (1978) demonstrates the high rates of helminthic and parasitic infections in pre-school children in a slum area of Delhi, while Sabir (1984) found that 73 per cent of the children in 151 households of a Lahore slum were malnourished and 61 per cent were growth stunted. In a study in a large slum complex in Bombay, 10 per cent of children under five years of age were vitamin A deficient, 20 per cent were suffering from protein–energy malnutrition, and 30 per cent from rickets (Jha 1985).

In another study of a large slum in Bombay (Desai and Pillai 1972), half the families surveyed reported some major illness during the previous year. The most common complaint was diarrhoea and other gastrointestinal dis-

Fig. 4.3 Children are at particularly high risk and are often exposed to the cycle of infectious disease, malnutrition, and mortality. Photograph by J. Holland, OXFAM.

Table 4.1 Most frequent diseases in children from birth to 6 years old, Coatzacoalcos, Mexico, 1984–5

(1) Diarrhoea and dystentery
(2) Influenza and coughs
(3) Bronchial infections
(4) Asthma
(5) Tuberculosis

Number of children in survey was less than 240.
Source: Unpublished data from a community survey (1983).
For details of the health programme see Bosnjak and Cayon (1985).

Table 4.2 Immunization coverage of children aged 6 years in Coatzacoalcos, Mexico, 1984–5

	BCG	DPT	Polio	Measles*
August 1984	60.2%	43.6%	44.1%	49.2%
December 1984	67.5%	54.6%	53.3%	58.8%
May 1985	76.3%	61.3%	65.0%	76.0%

*Children from 12 to 60 months.
Source: Unpublished data from a community survey (1983) For details of the health programme see Bosnjak and Cayon (1985).

orders, and 1 in 10 reported tuberculosis. Environmental contamination in Jakarta was measured in one study (Gracey *et al.* 1976) using bacterial cultures from a river running through a residential area, from open drains, roadside puddles, riverside wells, and ice-lollies sold to children in the street. A large number of micro-organisms were found, including pathogens like *Salmonella, Shigella*, and *E. coli*. In some specimens the river water contained a density of faecal bacteria of the same order as the human intestine. It is probable that such heavy environmental contamination is partly responsible for the high incidence of diarrhoeal diseases.

As opposed to special 'studies', health programmes which have good evaluations can also provide us with detailed pictures of the health status in low-income urban areas. For example, Tables 4.1, 4.2 and 4.3. provide information about the health status of residents of a shanty town in Mexico, including immunization coverage.

Intra-urban differentials in health

The study of intra-urban differentials, these are the differences that exist between rich and poor, is in its infancy. People seldom realize that there are

Table 4.3 Percentage of population with intestinal parasites in Coatzacoalcos, Mexico, 1983

	%
Trichuris (whipworm)	79.8
Ascaris L. (roundworm)	68.1
Uncinaria	27.8
H. nana	14.2
Giardia L.	9.5
E. histolytica (amoebic dysentery)	5.1
Strongyloides	2.8
H. diminuta	0.9
Taenia sp. (tapeworm)	0.2
Population:	
Without parasites	9.3
With parasites	90.7
Of those with parasites:	
Single infections	13.7
Multiple infections	86.2

Source: Unpublished data from a community survey (1983). For details of the health programme see Bosnjak and Cayon (1985).

> urban groups whose health conditions are in several ways worse than those of corresponding rural groups. Seldom do there exist in the rural areas the appalling conditions of extreme misery, destitution, environmental degradation and moral disruption that affect huge populations in many large and intermediate cities of the developing world. Without wasting resources or burdening the system with unnecessary data, an effort must be made to delineate *high risk* areas and population groups, and to provide minimum, purposeful, properly disaggregated information, that is sufficient to identify and illuminate the problems and to monitor change. (Rossi-Espagnet 1984, p. 42.)

City health statistics usually tend to look much better than rural ones. Basta (1977) has suggested that the reason is either because the squatter or slum inhabitants are not included in the statistics (they are not 'official' residents of the city in many cases), or because their inclusion is obscured by the enormous differences that exist between their status and that of the small, middle- to high-income parts of the city. Thus, a very misleading average often becomes the basis of city statistics. Properly compiled and disaggregated information reveals a quite different and more truthful picture. Surprisingly, however, as Rossi-Espagnet (1984, p. 14) has stated 'A systematic study of intra-urban differentials in health and health-related conditions has not been carried out anywhere in the developing world.'

Table 4.4 Differences among infants and young children in Manila and a squatter settlement (Tondo), Philippines, 1975

Indicator	Manila	Tondo
Moderate protein–energy malnutrition (%)	21	37.5
Severe malnutrition (%)	3	9.6
Neonatal mortality/1000 live births	40	105.0
Infant mortality/100 live births	76	210.0

Source: Basta (1977).

It will be clear by now that data which compares different social strata, income groups, or geographical areas in any one city are very rare. However, a few references do highlight intra-urban differences and comparisons between the urban poor and rural poor.

The most useful piece of work to date on intra-urban differentials is that by Basta (1977). Drawing upon several projects, Basta points out that many of the major cities of the developing world report infant mortality rates (IMRs) of between 75 and 90 per thousand births, but amongst the urban poor these rates are far higher. In Manila, the amount of severe malnutrition and the IMR is three times higher in the slums than in the rest of the city (see Table 4.4). The rates for tuberculosis were nine times higher, and diarrhoea twice as common as in the rest of the city. Twice as many people were found to be anaemic and three times as many were suffering from malnutrition compared to the rest of the city. Again, in the *bustees* of New Delhi the overall child mortality rate (0–5 years) is 221/1000 but reaches twice that number amongst certain castes.

Table 4.5 shows the differences in infant mortality rate both between slums and a middle-class area and between ethnic groups within the slums of Karachi, Pakistan. Such a useful breakdown of information is rare.

Other work which analyses intra-urban differentials includes that by Bianco (1983) who shows that mortality by tuberculosis is three times higher in the peripheral areas of Buenos Aires than in the central city. Cassim *et al.* (1982) indicate that the IMR in Colombo is significantly higher in the shanty-towns than in higher-income areas.

Other relevant studies are noted below:

- Bombay. In one slum the overall prevalence rate for leprosy was 22/1000, compared to a city average of 6.9/1000 (Ganapati *et al.* 1976; Ganapati 1983). The island city's crude death rate is twice as high as that of the suburbs and almost three times that of the extended suburbs: crude rates of 130, 73, and 54 per 1000 respectively in 1980–1 (Desai and Pillai 1972; Ramasubban and Crook 1985).

Table 4.5   Intra-urban and intra-slum differentials in infant mortality
rates in Karachi, Pakistan, 1985

Area	Infant mortality rate	Deaths of under-fives as % of all deaths
Orangi[1]	110	51
Chanesar Goth[1]	95	50
Muslim	77	
Hindu	146	
Grax[1]	152	55
Christian	123	
Muslim	139	
Hindu	286	
Karimabad[2]	32	18

[1]A *katchi abadi* (slum).
[2]Middle-class area.
Source: Agha Khan University students, 1986 (unpublished).

- Singapore. Incidence of hookworm, *Ascaris*, and *Trichuris* was 75.4 per cent in squatters, compared to 32.1 per cent in flat-dwellers (Kleevens 1966).
- Panama City. Of 1819 infants with diarrhoeal diseases, 45.5 per cent came from the slums, 22.5 per cent came from the shanties and none from those children living in better housing (Kouray and Vasquez 1979).
- Bogota. The southern and peripheral areas of the city have a much higher incidence of malnutrition as compared with the rest of Bogota (Mohan *et al.* 1981).
- Delhi. Rates of vitamin A, B, and D deficiency and malnutrition were higher in pre-school children who came from the slums (Datta Banik 1977).
- Sao Paulo. Infant mortality rates varied from 42 per thousand in the core areas to 175 in one of the peri-urban municipalities. Infectious diseases accounted for one-third of all infant deaths in the core areas and almost one-half of infant deaths in the periphery. Infant deaths rates for enteritis, diarrhoea, and pneumonia in the periphery were similar (about 1200 per 100 000 population) and are twice as high as in the core area. Neonatal deaths predominated over late infant deaths in the more affluent areas (core and intermediate), while the reverse has been true for the periphery of Sao Paulo. While cardiovascular diseases remain the leading cause of death in all three areas, they make up 20 per cent of total deaths in the core areas and only 10 per cent in the periphery (World Bank 1984).
- Porto Alegre. Similar differentials to those in Sao Paulo were found. In 1980 one out of every five infants in Porto Alegre was a shantytown resident. The infant mortality among the shantytown residents (75.5 per thousand) was three times as high as among the non-shantytown residents (24.4 per thousand) in 1980. As in Sao Paulo, there is a predominance of post-neonatal over neonatal mortality in the shantytowns, and vice versa in the non-shantytowns (see

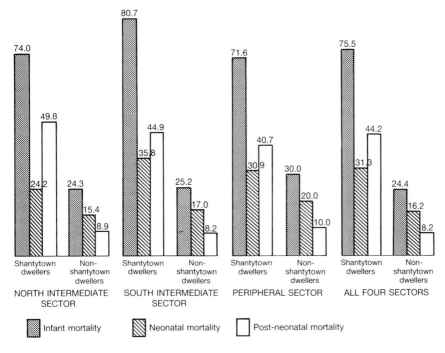

Fig. 4.4 Infant mortality (deaths per thousand infants), neonatal, and post-neonatal mortality among shantytown residents and non-shantytown residents in the Porto Alegre study area and in three of the four study sectors, 1980. *Source*: Guimaraes and Fischmann (1985).

Fig. 4.4). This is presumably due to the greater exposure of shantytown infants to communicable diseases and adverse living conditions. In this latter study, 64.5 per cent of the deaths among non-shantytown infants were caused by problems of gestation, delivery, or the puerperium and only 25.3 per cent were caused by pneumonia, influenza, infectious intestinal diseases, or septicaemia. In contrast, 50.8 per cent of the deaths among shantytown infants were attributed to these latter causes. At the same time, the mortality from most of these causes was substantially higher in the shantytown population (e.g. the mortality from pneumonia and influenza was six times higher, and that from septicaemia was eight times higher) than in the non-shantytown population (Guimaraes and Fischmann 1985).

One aspect which is rarely considered in the literature is the difference in health between poor urban communities within a city. Some exceptions are Bapat and Crook (1984) who examine the environment, health, and nutrition in different types of settlements in Poona, and Schensul (personal communication) who is developing a typology of poor urban communities in Lima based on 'urban services' and 'social organization'. Preliminary results from Lima indicate that diarrhoea and respiratory problems may be associated with the 'less developed' communities (i.e. those with less urban services and less social organization) while malnutrition may be associated with the more developed communities. It is suggested that the more developed stage demands more of the family budget in building permanent housing and contributing to co-operative water, sewage, and electrification projects. Scarce cash which goes to these essential resources may mean less food for children in the family. Exactly the same kind of results were discovered in the 1930s in Britain when slum dwellers were moved into new housing (M'Gonigle 1933). Further details of this study are given in Chapter 6.

Comparisons of urban and rural populations

A survey analysis in India as long ago as 1958 found that 'with regard to sickness . . . rates were somewhat higher in urban communities than in rural areas; in the latter, the highest and lowest incidence rates were 49 per 1000 population and 26 per 1000 as compared with a high of 55 per 1000 and a low incidence rate of 30 per 1000 population in urban India . . . This evidence of higher sickness rates for urban India conforms with the higher mortality rates among urban inhabitants.' (Johnson 1964.) The following discussion on urban–rural differentials in health is divided into that on nutritional status, infant and child mortality, and on other health problems.

Nutritional status

After examining available literature Austin (1980) and Lee and Furst (1980) concluded that problems of substandard nutritional status are more prevalent in urban areas than in rural areas in developing countries, especially among the lower socio-economic segments of urban populations. 'This is expected because of greater monetization of urban economies than of rural economies.' (Lee and Furst 1980, p. 10.)

Studies which address comparisons between the health of urban poor and rural populations have usually found that there were more severely malnourished children in low-income urban than in rural populations (Monckeberg 1968; Khanjanasthiti and Wray 1974; Prasada Rao *et al*. 1974; Stephens 1976; Arias 1977; Kerejan and Kowan 1981; Brink *et al*. 1983; Sudharto 1983; Diskett 1986). For example, Fig. 4.5 shows the higher prevalence of malnutrition in an urban slum compared to a rural village in Thailand, with severe protein–calorie malnutrition for children under six months old in the slum. Nelson and Mandl (1978) have analysed the reasons for this. Although in South-East Asia and Latin America rural labourers largely depend on their landlords for their food, many rural families, especially in Africa, own a small piece of land where they can grow part of their

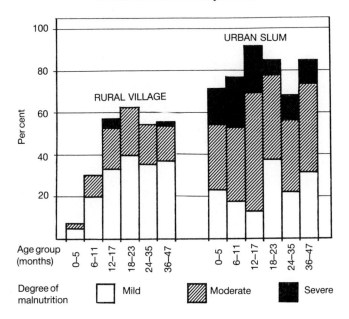

Fig. 4.5 Prevalence of malnutrition among rural and urban children by age and severity, Thailand, 1972. *Source*: Wray (1978).

food, or where harvest surpluses are available: this is generally not possible for the poor in the overcrowded cities.

In the cities, although salaries are higher, so also are costs, with the result that the poor have a smaller proportion of their income available for food. Furthermore, in the highly competitive situations of the city, women are often forced to work in full- or part-time jobs (generally in the informal sector) to complement the family income or as the only family support. Under such circumstances, women may typically have less time for food preparation and they may resort to early weaning or often not breast-feeding at all. They may leave their infants in the custody of young children who are unable to prepare weaning food properly. They may have to dilute and divide a limited milk supply among many infants. They may also fall easy prey to advertisements for breast milk substitutes (Plank and Milanesi 1973). (See Ch. 6 for a full discussion of malnutrition in the urban context).

A detailed study of the levels of food expenditures and consumption in urban and rural areas of Indonesia was carried out by Chernichovsky and Meesook (1985). They found that in spite of the relative affluence of the urban population, it does not fare better than the rural population in terms of diet. Urban diets are more expensive; relative prices also bias consumption away from grains which are rich in carbohydrates towards other foods which are rich in protein and fat. The urban population is, on average,

Fig. 4.6 Mothers often work to complement the family income or to provide the only family support. Infants often have to be left in the custody of young children unable to prepare weaning food properly. Photograph by V. Virkud.

better-off in terms of the consumption of protein and vitamin C and worse-off in terms of calories and micronutrients. The authors conclude that because of a more severe distributional problem in urban areas, the urban population may be more at risk of malnutrition than the rural population, especially if incomes or prices change adversely. They suggest that this situation may represent one of the more basic problems associated with urbanization in Indonesia.

Other rural–urban comparisons of per capita food consumptions showing a similar pattern are presented in Table 4.6, developed by Basta (1977). In all but one country, Tunisia, rural consumption levels were higher than those of urban residents. Furthermore, Basta contends that intra-urban differences are even greater than urban–rural differences and therefore, that the comparison of averages are misleading, especially for urban groups where wide disparities in income would push average calorie consumption per person much lower for poor income groups in cities. To illustrate the range of disparity in nutrition status Lee and Furst (1980) presented Brazil's and Sri Lanka's data differentiated by income levels for urban and rural populations. Calorie intake data from three regions of Brazil stratified by income for urban and rural areas indicated that for the lowest income groups in all three areas, urban intake was lower than rural. Even in Sri Lanka in 1982, where calorie and protein consumption on the average appeared to be ade-

Table 4.6 Intake of selected nutrients and energy in urban and rural areas of nine developing countries, 1960–70 (Amounts per head per day)

		Energy (keal)	Protein (g)			Fat (g)
			Total	Animal	Vegetable	
Tunisia 1965–7	rural	2315	63.7	7.4	56.3	55.2
	urban	2550	67.7	15.0	52.7	77.5
Trinidad and Tobago 1970	rural	3011	81.7	31.8	49.9	84.0
	urban	2550	83.6	43.3	40.3	95.8
Chad 1965	rural	2467	90.1	10.5	79.6	62.9
	urban	2113	73.3	23.9	49.5	52.2
Benin 1966–7	rural	2142	51.0	7.0	44.0	47.4
	urban	1908	52.0	10.0	42.0	46.2
Morocco 1970–1	rural	2888	84.0	12.0	72.0	*
	urban	2521	70.0	19.0	51.0	*
Brazil 1960	rural	2640	79.2	29.7	49.5	60.0
	urban	2428	74.0	30.7	43.3	63.0
Bangladesh 1962–3	rural	2254	57.4	7.9	49.5	17.2
	urban	1732	49.5	12.1	37.4	25.0
Pakistan 1965–6	rural	2126	69.8	7.9	61.9	40.3
	urban	1806	58.4	9.8	48.6	40.0
Republic of Korea 1969	rural	2181	66.9	5.1	61.8	15.8
	urban	1946	62.8	10.9	51.9	19.5

*Not available.
Source: Basta (1977).

quate for both rural and urban dwellers, consumption was consistently higher for rural inhabitants (see Table 4.7).

In terms of nutrition status, we conclude that the idea that health conditions are worse in rural compared to urban areas is a myth. Often, malnutrition is more prevalent in urban areas than rural areas and the extent of malnutrition among the urban poor is sometimes more severe than among their rural counterparts.

maln
urban >> rural

Infant and child mortality

The general impression that the urban population has an advantage over

Table 4.7 Apparent per capita nutrient available, by income groups in Sri Lanka, 1972

Income (in US$)	Urban		Rural	
	Calories	Protein (g)	Calories	Protein (g)
Below 200	1901.5	44.2	2099.0	46.5
200–399	2067.0	47.7	2326.4	52.5
400–599	2230.4	41.9	2466.9	56.0
600–799	2339.7	54.8	2598.2	59.0
800–999	2450.1	59.1	2582.2	60.3
1000+	2496.3	62.2	2888.8	67.4

as Y ↑ so does kCal

Source: Perera (1972).

Table 4.8 Death rates of 0–4-year-olds in six Latin American countries for 18 urban and rural sample areas (per thousand population)

	Urban	Rural
Argentina		
Chaco	20.7	23.9
San Juan	12.9	24.0
San Juan Suburban	21.9	
Bolivia		
La Paz	26.6	
Viacha		48.1
Brazil		
Ribeirao Preto	10.9	13.0
Recife	29.3	
Franca	19.4	
Sao Paulo	17.7	
Chile		
Santiago	13.0	14.0
El Salvador		
San Salvador	26.4	50.5
Jamaica		
Metro Kingston	10.5	
St. Andrews		9.5

Source: Lee and Furst (1980), reprinted from Puffer and Serrano (1973).

rural, as revealed by aggregate mortality statistics also needs modification. Death rates for those under five years of age in six Latin American countries have been estimated by Puffer and Serrano (1973) for both urban and rural samples (Table 4.8).

Of the six countries observed, only two countries, Bolivia and El Salvador were found to have rural childhood mortality clearly higher than the urban childhood mortality.

In Brazil, de Carvalho and Wood (1978) found that the life expectancy of the urban poor of Sao Paulo and Belo Horizonte is worse than that of low-income rural households. On the other hand, mortality rates for upper income urban groups are better than those of their rural counterparts.

In the large slum areas of Port-au-Prince (Haiti) over 20 per cent of newborns die before one year of age and another 10 per cent or more succumb in the second year; these mortality rates are almost three times those of the rural areas (Rodhe 1983).

Lee and Furst (1980, p. 25) conclude in their own summary of infant and child mortality statistics that 'the high mortality rates of the lower socio-economic class of urban populations are more comparable to their rural counterparts than their fellow urban dwellers of the upper class.'

Other health problems

Other urban–rural differentials are highlighted by Coulibaly (1981), who found that the average annual incidence of tuberculosis infection in the Ivory Coast was 1.5 per cent, and that this covered a range from 0.5 per cent in the rural areas to 2.5 per cent in Abidjan. In the more deprived areas of Abidjan (Vridi and Koumani) the prevalence could reach 3 per cent and the disease could also be present at much younger ages. A similar pattern of results has been found in Bangladesh (Ishikawa and Nabi 1981). A similar differential was observed in nineteenth-century England.

In Dakar (Benyoussef *et al.* 1973) one-third of a peri-urban sample was positive for *Ascaris*, while only three cases in a sample of 400 were found in the rural area. Similarly, Richardson *et al.* (1968) found that *Ascaris* prevalence in Dube, Soweto, was seven times that in rural 'highveld' communities.

Essentially, in the measurement of health conditions socio-economic class seems to be a more discriminating factor than rural-urban residence. As Lee and Furst (1980) have emphasized, it is therefore of the greatest importance to look beyond the rural–urban dichotomy of poverty as indicated by various living conditions, and toward those socio-economic factors which transcend geographical location.

Summary

- The health problems of the urban poor can be categorized into direct problems of poverty, the physical environment, and psycho-social issues.
- The urban poor suffer health problems of 'underdevelopment', such as malnutrition and infectious diseases, *and* those of industrialization, such as the chronic and social diseases.

- City health statistics tend to look much better than rural ones because either the urban poor are excluded or because their inclusion is obscured by the enormous differences that exist between their status and that of higher income groups. There is a need for properly disaggregated data to clearly demonstrate the full differentials in urban health.
- When such data are examined, the health conditions of the urban poor often appear as bad or worse than those of their rural counterparts, especially in terms of nutrition.
- There is also a need for more information which highlights the differences that exist in health status between and within poor urban communities in a city.

Fig. 4.7   Active in the city. *Source*: D. Brunner.

5

Primary health care and the urban poor

> So that our efforts do not become mere palliatives reinforcing the unjust structures that perpetuate poor health services, health should be viewed as inter-related with the problems of unemployment, high prices, inadequate housing, etc. Health care, be it in an urban or rural setting, to be liberating in action for the poor, should take into consideration the root causes of ill-health.
>
> Jaime Galvez Tan, (1985, p. 2).

The problem

As the previous chapter has indicated, the health problems of the urban poor have poverty as their root cause. Poverty as expressed by lack of food, lack of finance, lack of education, poor sanitation, and inadequate housing is a major cause of the health problems described earlier. Changes in these basic causes of ill health will do as much, if not more, to improve health than health services will. In this chapter we examine the problem of the current, mainly curative, health services for the urban poor and argue that there is a need to focus on more preventive aspects of health care.

In spite of the well-known concentration of health resources in the cities compared to the rural areas, and the relative proximity of hospitals and other medical facilities, for those who live in the slums and shanty towns of the developing world standards of health services fall far below reasonable minimum levels. Lack of care is far graver than each city's overall mortality and morbidity data suggest, for these data are, as already mentioned, averages, concealing the large differences between the best and the worst figures. Some have spoken of an 'inverse care law', whereby those in greatest need of medical care have the poorest access to it.

City administrations can hardly keep pace with the scale and tempo of urbanization and the multiplicity of problems that go with it. Health and other social services already existing in the city are not equitably distributed, nor are they planned, designed, or implemented to help those who are, in fact, in greatest need. Yet, poorer people contribute to the cost of these services through various forms of direct and indirect taxation, and through a variety of additional costs imposed on them by the location and operation of the services concerned (Rossi-Espagnet 1985).

The inadequacy of health services and health delivery is echoed by many writers. Teller (1981) points out that if the urban poor in Latin America fail

57

Fig. 5.1 Although spatial access to health services is equal for these low-income and middle-income areas, the services are socially and financially inaccessible to the poor. WHO photograph by Takahara.

to utilize modern treatment facilities it is because the staff disregards their most cherished values: personal respect and individual consideration. Similar remarks are made about the urban poor's under-utilization of health care in Seoul, Korea (Chung 1980). In Jakarta, Indonesia, it was discovered that most clinics, even in areas of high disease prevalence, were little used by the local population—poor services, crowded waiting rooms, and socially distant personnel being at least partly responsible (World Bank 1978). Many surprising situations exist. In a plan for the development of health services in The Gambia, the peri-urban area of Banjul was identified as the least served area in the country (Wheeler, personal communication). This situation co-exists with the substantial concentration of hospital services in the city itself.

Ongari and Schroeder (1985) describe the situation in slums around Nairobi:

> 'Looking at the health facilities around the slum areas, we see that they are not enough for the people living in the slums. That means that people have to queue, often half a day or more. Let us take for example a dispensary, where people queue half a day. Then they face the problem that there are no medicines. So now they are waiting for nothing, getting told to go to the chemist to buy the medicine. But that is something they cannot afford. To underline this

fact we quote the Sunday Standard from 21st April 1985: *Medical services at the Kenyatta National Hospital Nairobi are coming to a standstill following an acute shortage of a wide range of drugs, particularly antibiotics. An increasing number of both in and out-patients at the hospital, most of whom are poor people, are being forced to buy prescriptions from private chemists, while others are referred to private hospitals for such simple services as blood tests*. . . . If the mothers go to the clinic for family planning, infant welfare and child welfare, then she needs to go three mornings, because the clinic staff do one day family planning, another day infant welfare and a third day child welfare. There is of course queuing again, and so this mother spends again three half days. When shall she actually earn her living? Another problem in the health facilities is that the slum dwellers are often treated in a very rough way. The reason for that is, that the facilities are often under-staffed, so that the nurses are over worked and tired and they do not explain to their patients what their problem is. Secondly they are frustrated when they want to give treatment but there is no medicine. On top of this, unfortunately some nurses or clinical staff members feel superior towards the slum dwellers. All these factors do nothing to encourage the people in the slums to go to the health facilities for their services.'

A similar situation exists in Bombay, described below by Parikh (1985). The existing health services fall short of requirements and are managed in such a way that practically nothing reaches the poor—the most needy.

The health care system is mainly curative and is carried out by both the private and public sectors. There are hundreds of private dispensaries conducted by trained, half-trained, or untrained people of every available system in the world, starting from occult to most modern medicine. Similarly, there are specialists in every area of medicine and in every system—Bombay boasts of the best 'medical care' in India. No data is available from the private sector as there is rarely a system of record keeping in private practices. Even the notifiable diseases are hardly notified and, unless they cannot be treated at home, they are not referred to hospitals. On a first visit even a poor patient spends anything between 10–100 rupees, according to the seriousness of the condition. The money is usually borrowed or something in the house has been sold or mortgaged. The next day there is hardly any money left for sustained treatment.

A large number of private registered practitioners have what is known as an ESIS (Employment State Insurance Service) practice. The scheme is for industrial workers and their families who are permanently employed full-time, and getting a monthly salary of Rs. 600 or less. Many workers do not benefit from this scheme. For the Central Government servants there is a similar scheme which is also in the same sorry state. Unfortunately, both the schemes, though good on paper, are misused both by the workers and the doctors. There are hundreds of private nursing homes and maternity hospitals mostly for the middle class who can afford the high charges. There are quite a few sophisticated hospitals run by charitable trusts or societies but these are beyond the reach of the majority of the population.

In the public sector there are 150 municipal dispensaries supposedly catering to the daily needs of the poor people. In addition to these, every public hospital (governmental or municipal) has an out-patient department attempting to serve the general population. These dispensaries and out-patient departments are entrusted with both curative and preventive health services. They are not geographically well distributed throughout the city nor are they managed in a very efficient way. Many of the dispensaries are ill-equipped to serve the patients. They are often overcrowded and under-staffed. There is no regular referral system with any of the hospitals. No follow-up is possible of the chronically ill patients who cannot afford to go to private doctors. They just get worse at home and often only go to hospitals to die.

The number of dispensaries and hospitals is too small for the health needs of the population. There are hardly any health-promoting activities nor is preventive health emphasized. The Municipal School Children's health scheme is perhaps the only existing preventive health service. There is an experimental Integrated Child Development Scheme going on in two slums in the suburbs of Bombay.

As a final example, Fig. 5.2 demonstrates the inaccessibility of health services to a slum population in the city of Coatzacoalcos, Mexico (Bosnjak and Cayon 1985). The figure shows that people generally did not utilize or

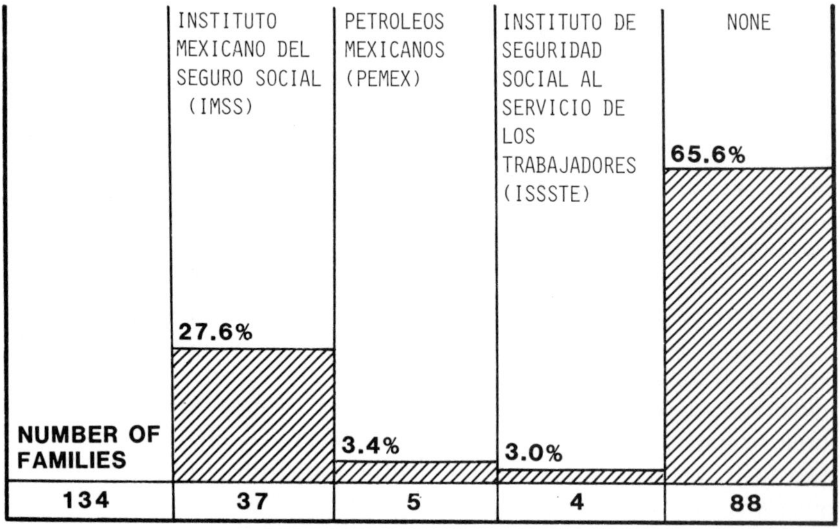

NUMBER OF FAMILIES	INSTITUTO MEXICANO DEL SEGURO SOCIAL (IMSS)	PETROLEOS MEXICANOS (PEMEX)	INSTITUTO DE SEGURIDAD SOCIAL AL SERVICIO DE LOS TRABAJADORES (ISSSTE)	NONE
	27.6%	3.4%	3.0%	65.6%
134	37	5	4	88

Fig. 5.2 Access to institutional health services, Coatzacoalcos, Mexico, 1983. *Source*: Bosnjak and Cayon (1985).

had no access to health services in the city even though a range of services existed (PEMEX serves employees, ISSSTE serves state employees, IMSS is for insured employees of the private sector). The government concentrates its provision of mobile health services in rural areas, neglecting poor urban areas.

Clearly, the current health services are failing to reach the urban poor. Even if the services *were* to reach these populations the emphasis upon curative services means that the underlying causes of ill health are unlikely to be tackled. What can be done?

A solution—primary health care PHC

Primary health care (PHC) concepts apply to urban health systems just as they do to the health systems of whole countries, for the discrepancy between the allocation of resources and health needs is equally striking in the cities, and the gaps between rich and poor, and between need and provision, are actually widening.

PHC represented a qualitative jump in relation to the old concept of basic health services. Its concepts convey, first of all, a philosophical message which emphasizes equity and justice in matters related to health. Secondly, it delineates a strategy that starts from an improved understanding of health problems (which often involves the political, economic, and social realities of each country) and attempts to find solutions that go beyond the technological treatment of diseases to their fundamental causes. Thus, the strategy also entails political decisions on matters such as employment and income, land distribution and tenure, basic education and housing; co-ordinated efforts by all the sectors concerned with socio-economic development; and a better balance between 'top-down' planning and the upward expression of needs, aspirations, and possible contributions by individuals and communities to their own development. Finally, PHC concepts emphasize action at the primary level, that is the level of first contact between the people and the system, particularly in relation to the eight tasks[1] proposed in the Alma Ata Declaration, with the rest of the system being responsive and supportive to action at the primary level and being conditioned by it. Near coverage is the essential target (Rossi-Espagnet 1985).

Since the Alma Ata Declaration in 1978, most of the initiatives in primary health care have been in rural areas. Programmes based on rural health needs, a rural social structure, and a rural administrative framework have now been in operation for several years. It is important to learn from these initiatives but what are the special problems that arise in the urban context? Are the processes involved different or the same?

More recently, the issue of comprehensive primary health care versus *selective* primary health care is being debated (see, for example, Rifkin and

DEBATE in PHC

Walt 1986). Some have argued that comprehensive primary health care as outlined above is idealistic and too expensive. Many programmes which use the rubric of primary health care are, in fact, selective primary health care programmes in that they focus upon selected components of PHC, such as immunization and oral rehydration therapy. Usually these components are technological medical interventions which have measurable or quantifiable results, are cost-effective, and maintain the international and national status quo (Unger and Killingsworth 1986). Proponents of comprehensive primary health care have argued that such selective programmes negate the concept of community participation, reinforce authoritarian attitudes, and have a questionable moral and ethical value in which foreign and elite interests overrule those of the majority of the people.

In the case studies in this book we will see a mixture of selective and comprehensive primary health care programmes. It is important to bear this distinction in mind when discussing any aspects of primary health care.

A policy that could be regarded as selective primary health care is the current 'Child Survival and Development Revolution' of UNICEF, which claims to enable parents to halve the rate of child deaths and save the lives of up to 20 000 children each day. As many programmes in poor urban environments are supported by UNICEF it is important to understand the components of this policy. Essentially, the emphasis is upon:

- **G**rowth monitoring through the use of growth charts on which serial weight for age readings are marked. It has been shown that these charts can help identify and draw attention to children at high risk from the synergistic effects of infection and malnutrition.
- **O**ral rehydration therapy (ORT) which is a safe salt and sugar solution effective in preventing death from dehydration caused by acute watery episodes of diarrhoea.
- **B**reast-feeding which, besides providing an ideal food, protects the infant from infection and acts as a natural contraception.
- **I**mmunization against the six preventable childhood infections (measles, whooping cough, tetanus, polio, diptheria, and tuberculosis).
- **F**ood supplements to malnourished children.
- **F**amily spacing through birth-control to improve the health of mother and child.
- **F**emale literacy which is known to be related to infant mortality and fertility rates, independent of socio-economic status.

These actions are usually summarized as 'GOBI-FFF'.

The use of different starting points

We have seen that the health problems of the urban poor relate to factors such as lack of food, income, education, and poor sanitation and housing, as well as health services. We have also discussed how the philosophy of the primary health care approach addresses such problems and emphasizes a preventive rather than a curative approach to health care.

In the rest of this book case studies are presented under various headings such as nutrition, income generation, community education, sanitation, and housing. This structure reflects the fact that many initiatives in poor urban communities which address health issues have an initial focus upon a main underlying cause of ill health. The programmes use different starting points. Thus, in the nutrition chapter we have case studies from Lima, Peru, and Port-au-Prince, Haiti, which address the problem of urban malnutrition in two different ways. Both case studies started by tackling malnutrition but both now encompass other components such as immunization, family planning, lobbying for improved sanitation, etc.

Two case studies from India are compared in the chapter on income generation. Both programmes tackle the problem of increasing the income of poor urban populations but one can be seen as an 'outsiders' attempt while the other was generated from within the slum. Both programmes now tackle health problems with specific interests in family planning.

Initiatives which realize that health education is needed before some poor urban populations feel the need for preventive health care are featured in Chapter 8. Both programmes discussed have ended up tackling a wide array of health-related problems.

Sanitation is a special problem in crowded slums and shanty towns. The two case studies in Chapter 9 demonstrate that sanitation is a useful entry point, in that communities were willing to organize themselves to improve sanitation and these communities then became interested in tackling other health problems.

Housing is often seen as being outside the realms of primary health care because of the cost of such infrastructure. However, self-help housing improvements cost little and can be linked to income-generating schemes (e.g. carpentry, mason skills). Also, the link between housing and health is poorly understood. The fact that improved housing is often a priority felt-need of the urban poor makes it an ideal starting point. The case study from Zambia in Chapter 10 shows how other activities related to nutrition and income generation can be linked into a self-help housing initiative.

Chapter 11 includes two case studies which have, from the outset, rejected the idea of focusing upon a single entry point and have taken a more integrated approach.

Finally, although 'health' in a narrow sense of the word is often not the top priority of the urban poor, there are programmes which use health services as a starting point. Chapter 12 includes case studies from Chile and Malaysia which had an immediate emphasis upon improving health services.

Note

1 Education concerning prevailing health problems and the methods of preventing and controlling them; promotion of food supply and proper nutrition; an adequate supply of safe water and basic sanitation; maternal and child health care including family planning; immunization against the major infectious diseases; prevention and control of locally endemic diseases; appropriate treatment of common diseases and injuries; and provision of essential drugs.

6

Nutrition

The need to buy food is one of the most important factors
responsible for high rates of disease and mortality among the
urban poor.

Mason and Stephens, 1981, p. 47.

Important factors

Urban malnutrition is fundamentally a manifestation of a larger syndrome,
urban poverty. None the less, since other factors are also important, the
multiple aetiology of urban malnutrition should be examined. Available
data on rural–urban and intra-urban differentials in malnutrition were
examined in Chapter 4. Here we examine some causes of urban malnutrition
which should be borne in mind when assessing the extent to which a programme is attacking a cause or a symptom. This section draws freely upon
Wray (1985).

Poverty

When total family diets reveal consumption levels substantially below minimum requirements, low income is the basic cause, especially when combined
with high unemployment, underemployment, or low wages.

Wray (1985) has drawn attention to a study that sheds provocative light on
the impact of poverty on nutrition and health in urban slums. The study was
carried out in a slum in Stockton-on-Tees, England (M'Gonigle 1933). In the
1930s the world-wide economic depression was perhaps at its worst, and the
situation of the people in that slum bordered on the desperate. Although
unemployment among heads of households was around 90 per cent, the town
council had obtained money to provide better housing. The medical officer
of health had the presence of mind to persuade the city officials to take
advantage of this and carry out a 'natural experiment'. The new and improved
housing that was built, described by M'Gonigle as 'everything that modern
sanitary science can demand', was better in every way in terms of space,
ventilation, water and sanitation. Following M'Gonigle's suggestions, the
authorities divided the slum and moved half of the families into the new
housing, leaving the remaining families to serve as a control. Thus M'Gonigle
was able, over a period of years, to estimate the impact of improved housing
on the health of slum populations.

M'Gonigle's results gave the mortality rates in the county as a whole, in
the city of Stockon-on-Tees, and in two parts of the original slum during two

periods: firstly the baseline from 1923 to 1927, and then between 1928 and 1932 when figures for the new housing area, Mount Pleasant, were added. The results astounded him and his associates. Crude mortality rates remained the same in the county and the city, but they had decreased by over 10 per cent in the original slum while they had increased by more than 30 per cent in the new housing area. This change, of course, was altogether counter-intuitive. M'Gonigle and his associates felt obligated to investigate it carefully, and, essentially, what they found was that the modest increases in rent that families paid for the new and better housing, came at the expense of their diet. Careful surveys of dietary intake among the families in the new housing area showed that their nutrition had suffered in comparison with that of families remaining in the slum, and that their reduced dietary intake seemed to have had a substantially greater impact on overall mortality than had the improvements in their housing. Similar contemporary results from developing countries were discussed in Chapter 4.

Lack of land on which to grow food

Malnutrition among the urban poor is partly a result of people being cut off from traditional opportunities to gather and cultivate food. The food intake of the urban poor often depends entirely upon their cash income, which can fluctuate wildly. Rural people may at least have a small plot of land on which to grow their own vegetables or rear some animals.

This need could partly be met through the introduction of home gardening schemes (see UNICEF 1984*a*). It is hypothesized that even a small ground space, roof area, window boxes, or trellises can enhance the prospect for reducing dependency of the urban poor on the cash food market. The value of increased nutritional intake for individual health, and also productivity, would also be important (Mason and Stephens 1981). Since 1978, owners of unused land in Manila are obliged to cultivate it or forfeit it to any other person willing to grow food on it. As a result a number of community gardens have been established. One of these supplied 800 squatter families with 80 per cent of their vegetables from an area of 1500 square metres (Wayburn 1985).

Infectious disease

Besides the lack of adequate diet, the urban poor are also likely to lack adequate water, sewerage, and health care. Congestion and crowding further exacerbate the health environment and increase the possibility of infectious diseases. Although such sicknesses and nutritional shortfalls may not be too severe individually, together their effect can be fatal, especially for young children.

The impact of various infectious diseases, such as measles, on nutritional status and mortality has been documented in several countries. In Haiti in

Fig. 6.1 The use of communal gardens such as this one in Kebele 41, Addis Ababa, Ethiopia, can improve diet. Photograph by C. Goyder.

1982, a prospective study was initiated in Cité Simone (now renamed Cité Soleil) on 600 6–12-month-old infants (Boulos 1985). Preliminary results suggest that measles has a significant impact on nutritional status. Children that developed measles were more likely to be severely malnourished and had a mortality rate that was three times as high as children who did not have measles. In this urban slum, measles transmission is high and occurs early: 40 per cent of children had already had measles by the age of 12 months. Besides measles, the most significant childhood infections are gastroenteric and respiratory diseases. The length and frequency of these attacks are major determinants of childhood malnutrition.

Decrease in breast-feeding

In 1972 nutrition surveys were carried out in several densely settled slum areas of Bangkok, and the results obtained in one of those areas are shown in the right half of Fig. 4.5. (Khanjansthiti and Wray 1974). Although it was anticipated that the situation of infants in the urban slums was poor, the dramatic differences in the nutritional status of those infants was unexpected. As the figure shows, not only were malnutrition levels very high, but it is clear that severe malnutrition was well established in the first six months of life. An explanation is that mothers in those slums, in common with slum-dwelling

Table 6.1 Proportions of infants breast-fed for 6 months or longer in
the total infant population, and among those infants dying at
6–11 months, in four urban study areas, c.1970

Study area	Total infant population		Infants dying aged 6–11 months	
	breast-fed for <6 months (%)	breast-fed for >6 months (%)	breast-fed for <6 months (%)	breast-fed for >6 months (%)
San Salvador, El Salvador	20	80	78.0	22.0
Kingston, Jamaica	51	49	87.4	12.6
Medellin, Columbia	61.8	38.2	91.3	8.7
Sao Paulo, Brazil	77.2	22.8	95.9	4.1

Source: Wray (1978).

mothers throughout the world, had to work, and so had to rely on artificial
feeding provided by some surrogate mothers or child minders.

Some hint of the impact of the presence or absence of breast-feeding on
the mortality of infants and young children in urban areas is provided in
Table 6.1. The table shows that mortality during the second six months of
life was much higher in those infants breast-fed less than six months than in
the infants that were breast-fed for more than six months.

Similarly, in the survey in the Haitian slum of Cité Simone mentioned
earlier (Boulos 1985), it was observed that infants that received any kind of
bottle feeding within the first month of life (56 per cent of infants) had
approximately a four-times greater risk of mortality than those that were
exclusively breast-fed throughout that period.

It was also shown that mortality rates by 36 months for children bottle-fed
in the first two days of life was 150 per thousand, compared to 23 per 1000
for children receiving their first bottle after the second day of life. A major
problem in urban areas is not merely shorter durations of breast-feeding, but
the very early introduction of the bottle in the infant's diet.

In Haiti the average duration of breast-feeding in rural areas is 18 months,
while in urban areas it is 12 months. This pattern is repeated throughout the
Third World. World Fertility Survey (1984) data (based on 42 developing
countries) found that rural women breast-feed for 2–6 months longer than
their urban counterparts. More substantial differences occur in Indonesia
(nine months), Jamaica (seven months), and Thailand (eleven months).
Other studies have found that less than five per cent of the women surveyed
in the cities of Sao Paulo (Brazil), Panama City (Panama), and San Salvador

(El Salvador) breast-fed for six months or more, and in the state of Sao Paulo, Brazil, less than 50 per cent breast-fed for as long as one month. The pressures from urban socialization that lead to a decrease in breast-feeding may also lead to other nutritionally harmful practices, such as acquiring high-cost and low-quality status foods.

LBW

Other factors—low birth-weight, short birth intervals, teenage pregnancies

In many instances low birth-weight (LBW) culminates in severe malnutrition during infancy and results in a greater susceptibility to infections, primarily due to low resistance (Shah 1983). In the short term, LBW increases the risk of neonatal death. This effect is not transient. In Cité Simone, Haiti, it was shown that children who had a low birth-weight had a mean weight of 7.1 kg at 12 months, while it was 9.1 kg for children whose birth-weight was more than 3330 g. In many poor urban areas there is an increasing percentage of these vulnerable LBW babies (see Fig. 6.2).

Child and maternal malnutrition is particularly common in large, closely spaced families. The phenomenon is accentuated when the mother delivers before her twentieth birthday. World Fertility Survey results substantiate this (Hobcraft 1985)—a child born less than 2 years after the previous birth experiences a 50 per cent increase in the risk of dying during childhood.

This excess risk persists across a wide range of mortality levels, broad

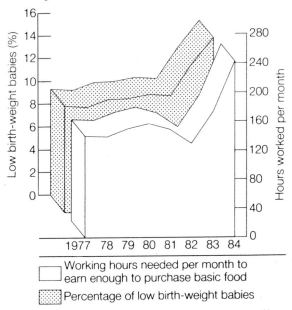

Fig. 6.2 Increase in low birth-weight, Recife, Brazil, 1977–84. 127 948 births (low birth-weight = less than 2500 g). *Source*: Dias *et al*. (1985).

geographic regions, socio-economic groups, parity, sex of the child, and familial levels of mortality. The magnitude of the differentials observed is comparable with the increase in survival chances associated with seven or more years of education for the mother over the child of an uneducated mother.

A World Bank study has estimated that elimination of all births occurring at intervals of less than two years might lower infant mortality rates by between 15 and 25 per cent in most Third World countries. These are significant potential gains and suggest the need for much greater emphasis on child-spacing in family planning programmes. Potts and Bhiwandiwala (1981) have provided a review of the issues involved in family planning for the urban poor.

Teenage pregnancies are increasingly common in urban areas of developing countries. It is now known that children born to teenage mothers are about 40 per cent more likely to die in the first year of life than those born to mothers in their twenties. These excess risks persist into childhood, being even greater for the second year of life. Moreover, the excess applies across all geographic areas (although slightly less for the Americas) and across different educational levels of the mother.

Case study 1 Cité Simone, Port-au-Prince, Haiti

Based on a paper by Boulos (1985).

Introduction

The Republic of Haiti had a total population of over 5 million people (according to the 1982 census) and, based on numerous indices of poverty, is classified as the poorest country in the Western Hemisphere. The average per capita income for 75–90 per cent of the population, i.e. the rural and urban poor, is US$150.00 per year. Port-au-Prince, the capital, contains approximately 1 million people. Large zones of poverty exist in Port-au-Prince where the standard of living is comparable to or, in certain respects lower, than that of rural areas. Cité Simone (or Cité Soleil as it is now known) is one of these 'peri-urban' slums.

Although there were few people in the area in the late 1950s, the arrival of rural migrants began and Cité Simone now has a population close to 100 000 inhabitants packed on less than 5 km² of land built entirely of landfill, a mere 1 m above the sea level, and bordering the bay of Port-au-Prince. Per capita incomes are very low, with living conditions marked by crowding and lack of sanitation. Some residents are relatively recent migrants who maintain close ties with their place of origin; others are second and third generation city dwellers, born and raised in marginal neighbourhoods of Port-au-Prince.

Formation of the 'Complexe'

In 1974, the Haitian Arab Centre, a private, non-profit-making group of Haitians, opened a small dispensary funded solely from local sources to provide free health care to Cité Simone residents. This group combined their efforts with those of other groups working to improve the living conditions of the residents, notably Foster Parents Plan and a group of Salesian priests. In July 1980, the *Centre Haitiano-Arabe* was incorporated under Haitian law with a congregation of Roman Catholic nuns, forming the *Complexe Medico-Social de la Cité Simone.*

At the earliest stage of its development, the directors of the *'Complexe'* felt that only a preventive approach could help improve the health conditions of the poor living in Cité Simone. An outreach programme was implemented using volunteers from the community as health collaborators, with a ratio of one for 500 people. Special services such as a dental clinic, TB clinic, and a laboratory were added as the programme developed. In 1981, a 70-bed general hospital opened to receive patient referrals from the different clinics of the *Complexe*. Finally, in 1983, a research and evaluation section was added to the *Complexe* in order to monitor and evaluate the impact that the different interventions were having on the health status of the residents of Cité Simone.

goal

The *Complexe* now consists of four principal medical facilities; the original Haitian–Arab Health Centre, which is an out-patient clinic; two medico-social centres in the Cité Simone neighbourhoods known as 'Brooklyn' and 'Boston', where preventive health services are combined with remedial education and job training activities; and the 70-bed general hospital, with laboratory and radiology services.

via: med services

In addition, the *Complexe* operates neighbourhood family nutrition demonstration centres scattered throughout the urban slum. Nutrition auxiliaries staff these centres, providing daily classes in food preparation, sanitation, and home economics for young mothers.

h ed

The *Complexe* also manages a wide variety of educational and social services including the Boston Cultural Centre, which offers remedial education and handicrafts skills and training for approximately 300 adolescents; the Brooklyn Mothercraft Centre, providing some 250 mothers of severely malnourished children the chance to learn sewing and handicrafts while their children receive treatment; and the *'Centre Papa Yo'*, a vocational and technical school, where fathers may train for jobs in local industry.

educ & soc services

The Complexe *and the challenge of malnutrition*

The approach to health care delivery in Cité Simone has changed over time from offering services through dispensaries; through setting up an outreach

Fig. 6.3   A nutrition education meeting in Dakar, Senegal, is one of the methods for combating poor nutrition in the slums and shantytowns. UNICEF photograph by S. Sprague.

programme with 112 Community Health Workers (CHWs); to health interventions targeted to particular high risk populations. This approach to service delivery developed after examining the results of a 1981 sample survey. The *Complexe* recruited more staff with public health training and an interest in action-oriented research designed to improve the health status through low cost, effective measures.

Malnutrition has remained a challenge for the *Complexe*. It is well recognized that malnutrition carries with it a long list of detrimental consequences. In Cité Simone, the risk of dying for a child admitted to hospital when suffering from third degree malnutrition is three to four times higher than for a child with mild or moderate malnutrition.

The combination of severe malnutrition and diarrhoea leads to extremely high mortality. Because of the unsanitary conditions prevailing in the urban slum, the incidence of diarrhoea is very high. In Cité Simone, on average, every child has one episode of diarrhoea per month and the principal cause of death for the under-fives admitted to St. Catherine's hospital, Port-au-Prince, in 1983 was gastroenteritis. Children suffering from malnutrition are at greater risk of dying from these episodes of diarrhoea and they contribute disproportionally to the child mortality.

This is further demonstrated by Augustin (quoted in Wray 1985) who

hi t in children

found that 67 per cent of the infant mortality in Cité Simone, was due to diarrhoeal disease, that 63 per cent of that mortality occurred in infants with severe malnutrition, but that only 3 per cent of the population was classified as having severe malnutrition. Thus, roughly half of the infant mortality was occurring in a highly vulnerable 3 per cent of the population.

For the past seven years, the nutritional surveillance programme allowed a quick identification of the severely malnourished children. CHWs are assigned a specific day in a three-month cycle on which they invite all mothers and children under five years old to the well-baby clinic. Weight is taken and plotted on the 'Road to Health' card and the nutritional status is defined according to the Gomez classification. Mothers learn to understand the card and keep it with them at home. This card is regarded as the birth certificate of the child. Children with third degree malnutrition are then referred to the nutritional recuperation centre for a period of approximately three months. All children suffering from mild to moderate malnutrition receive a card which allows them to receive a weekly food supplement. During their visit to the clinic, mothers of children with any form of malnutrition receive special attention from the public health nurse in charge of the nutritional surveillance programme, who provides health education and care to all mothers of children seen at the clinic. However, things have not improved as hoped.

Malnutrition prevails

Table 6.2 provides data on the nutritional status according to the Gomez classification (based on weight for age) of children under five in Haiti and Cité Simone in 1978. The Cité Simone data came from a special study conducted in January–March among 4000 children under nutritional

Table 6.2 Nutritional status of children under 5 years of age according to the Gomez classification (weight for age) in rural areas, Port-au-Prince, and Cité Simone, Haiti, 1978

| Area | Normal nutritional status (%) | Malnutrition | | |
		1st Degree (%)	2nd Degree (%)	3rd Degree (%)
Rural	24.1	46.4	26.0	3.5
Port-au-Prince	41.7	43.8	13.1	1.5
Haiti	26.8	46.0	24.1	3.2
Cité Simone	35.0	43.3	17.4	3.3

Source: Haiti Nutrition Status Survey (1978, p. 28) and unpublished data from Cité Simone (quoted in Boulos 1985).

surveillance. The other data are from the National Nutrition Status Survey. In 1978, only about one-third of the under-fives in Cité Simone were normal and 3.3 per cent were severely malnourished. Surveys in 1983 and 1984, based on samples of 12 000 and 16 000 children under surveillance, respectively, suggested continuing problems. Over 25 per cent of the children still suffered from second or third degree malnutrition. Only one-third were normal. The prevalence of severe malnutrition in Cité Simone, in spite of recuperation efforts and nutrition surveillance, has remained practically the same. Some possible explanations include:

- The fact that infant mortality has dropped considerably means that an increasing number of children with severe malnutrition, who would have died otherwise, are being saved by the intensive neonatal and infant care provided since the opening of the hospital.
- The high migration rate contributes to the cohort of children coming from rural areas where the prevalence of severe malnutrition is higher. These migrating groups are likely to be the poorest of the poor.

Of course, the major obstacle to good nutritional status for children is the dismal economic situation of most Cité Simone residents. In addition, the generally unsanitary conditions, lead inevitably to high rates of infection and disease.

Fig. 6.4   Mothers in Cairo: the early introduction of unhygienic bottle-feeding and breast-milk substitutes increases the rate of child malnutrition in poor urban areas. UNICEF photograph by B.P. Wolff.

Fig. 6.5 Lack of breast-feeding is a cause of malnutrition in poor urban areas. A community health worker in Mathare Valley, Nairobi, Kenya, encourages a mother to breast-feed. Photograph by L.A. Ongari.

New interventions

While accepting these root causes, the project decided to look in greater detail at three contributing factors, and to seek to identify relevant interventions. These factors were low birth-weight, bottle-feeding, and measles. The project's findings on the relationship between bottle-feeding and infant mortality, as well as that of measles and malnutrition, were mentioned above. The programme's strong emphasis on health education has stressed the importance of breast-feeding and the danger of the bottle. It will, however, be some time before the impact of the work can be measured.

Long-term objectives of the project aim at reducing teenage fertility and the provision of better antenatal care, especially during the first pregnancy. Third trimester women who show signs of infection are being treated with erythromycin. This early treatment of infection is expected to affect birth weight significantly (Boulos 1985).

Future developments

In collaboration with the association of Private Health Institutions, a broad-scale effort to improve child survival was started in 1985. A major component will be the design, implementation, and evaluation of alternative methods of increasing child survival through:

- the reduction of the incidence of low birth weight infants;
- the increased utilization of family planning services, particularly among women at high risk of giving birth to a LBW infant;
- the increased participation of mothers in selective primary health care programmes such as oral rehydration therapy and immunization.

Case study 2 Pamplona Alta, Lima, Peru

Based on a paper by Creed (1985).

Introduction

According to the 1981 census, 65 per cent of the Peruvian population lives in urban areas. While the overall rate of growth is around 3.8 per cent per annum, that of shantytowns or *'pueblos jovenes'* is around 6 per cent. Increases in housing have not been able to match this growth. A recent survey in Pamplona Alta found an average of 2.6 persons per room. Moreover, 25 per cent of Lima's population have no access to running water and depend upon water lorries or bucket supplies.

Pamplona Alta is a *pueblo joven* situated on the hillsides of the desert area south of Lima. At the beginning of the project in 1975 Pamplona Alta had a population estimated to be around 70 000, divided into 16 sectors. The current population is estimated to be well over 100 000. In 1975 there was no running water, sewerage system, or rubbish collection, although there was electricity in most of the sectors. Water and sewerage systems were brought into the majority of the existing sectors during 1981. At the start of the project there were three health centres available for this population. In 1985 there were five health centres and four health 'modules' constructed in different sectors of Pamplona Alta. The health and nutrition programme began in two of these sectors.

The health and nutrition programme began after disillusionment with a Nutritional Rehabilitation Unit (NRU) that was set up in 1975. It became obvious that it was almost impossible to expect the mothers of the children in the NRU to participate in the preparation of their diets: there were, and are, too many pressures on mothers' time and energy, such as the needs of other children at home and, in some cases, the necessity to work away from the home due to the economic situation. Weekly group teaching sessions were subject to the same limitations and were not as successful as was hoped, due

Fig. 6.6 Parts of Pamplona Alta, Lima, Peru. The site of the second nutrition case study. Photograph by J. Holland, OXFAM.

to poor attendance on the part of the mothers. The most successful form of teaching was found to be the conversations between the mother and the nutritionist or nurse explaining the changes in the diet as the child recovered. This was seen by the mother to be the most relevant in terms of her child. The mothers brought their children to the NRU because they were ill and were seen by them to need primarily medical care and attention. The orientation and the atmosphere of the out-patient clinic was very much towards curing the sick child rather than the prevention of illness, and due to the severity of the malnutrition, nearly always accompanied by complicating infections, the NRU developed into a small hospital rather than a unit for nutrition rehabilitation.

As a result of this experience it was decided that in this environment the rehabilitation of the severely malnourished child should be initiated in a hospital, and that if preventive medicine and nutrition is to be carried out this really needs to be done in the community itself and not in a NRU.

An alternative—the community health programme

The work in the community was initiated as an outreach from the nutrition rehabilitation unit. NRU staff were invited by the leaders of the women's group of one sector of Pamplona Alta to give nutrition talks and demonstrations for the selection and preparation of low-cost, nutritious meals. In the course of these weekly meetings it was realized that learning about nutrition alone would not solve the nutrition and health problems of the community, and from this an integrated health programme was planned. The major health concerns of the women were the health and nutrition of their younger children, the health of the women, and nutrition in terms of the feeding of the family. The community leaders supported a growth-monitoring and well-child health check programme in two sectors, and facilitated the initial background health and nutrition status survey that was carried out.

The technical team

The technical team consists of a religious sister who lives in the area and is the prime co-ordinator with the community, a nutritionist who co-ordinates the programme, a senior paediatrician, two junior paediatricians, and a nursing auxiliary. Each member of the team works part-time in the programme, and the majority have participated in it since it started.

Growth monitoring and health meetings

The programme has always been based in the community itself, and the meetings for the children between 0–5 years of age are held monthly in each block (consisting of about 300 families) in a house offered by a community leader, a health delegate or neighbour in the block or, if more convenient for the community, in the community centre. At these meetings the health promoter weighs and measures the child and plots his or her growth on a 'Road to Health' growth chart kept by the mother. This is also used as an educational tool. The health promoter checks the immunization record of the child, and advises on the feeding of the child according to her age and nutritional status. In the case of diarrhoea or other illness, appropriate advice is given. A paediatrician is present in the block once every one or two months. Children under one year of age have a monthly check, those 1–2 years bi-monthly, and 2–5 years every three months. Children with a health or nutrition problem are followed more frequently.

Health promoters and delegates

During the early months of the programme certain women emerged as leaders and were interested in becoming health promoters in their communities. In some cases these were women who were already elected by their community as a health delegate or social worker. These women undertook courses in first aid and basic health in various institutions, including the local health centre, and in nutrition and teaching methodology with members of the technical team. By 1984 there were a total of 13 health promoters in three sectors of Pamplona Alta, including two *'ampliaciones'* (new squatter areas). The health programme also depends upon health delegates, who are the co-ordinators in their respective blocks; these have not had the training as health promoters but are important in the motivation of their communities.

The functions of the health promoters are multiple and include the evaluating and advising of children who come to the health check and the follow-up of those requiring special care. They are involved in the demon-stration of the preparation of oral rehydration fluids or appropriate food mixtures for a malnourished child. They provide first-aid care, carry out minor treatments, give injections, and run the immunization programme.

duties

At least once a year each area is re-surveyed by the health promoter or health delegate to ensure that all children under 5 years of age and pregnant women are invited to the programme.

Since 1981 the health promoters have run their own health and growth-monitoring meetings so that the visit of members of the technical team have become less frequent, allowing more time for them to develop pro-grammes in new areas. It had been anticipated that the attendance at these meetings would be less than when the paediatrician was present, but this was not the case, indicating that the community accepted this alternative care.

Although the health promoter is involved in these health services, she is seen primarily as a member of her community working towards an improve-ment of health conditions and a raising of health awareness, rather than being an extension of a government health service system. Initially, one or two of the health promoters saw themselves as having a higher status than the rest of the community and tended to adopt the hierarchical relationship they had observed in health posts. However, this was corrected as the programme evolved, in part through the reaction of the community itself.

role

During 1984 and 1985 the community leaders asked for the programme to be extended to work in three new '*ampliaciones*', and work has begun directly with elected health delegates of the community. Based on experience over the years, the programme is moving away from the more curative aspects of the health promoters' work to emphasize the motivating and educating function of the health promoter and delegates. In fact, the people have been using the curative services offered by the health promoters less and less over the years. Some of the health promoters have handed their work on to others, and those actively participating at the present time are involved in training the health leaders in the new sectors.

To pay or not to pay

None of the health promoters in the programme have been full-time, in fact all of them at one time or another have had jobs outside of the community. Due to the economic situation and the time commitments to the programme, it was decided a few years ago to give them a small remuneration. However, this has had the tendency to set them apart from the community rather than promoting their further integration. Consequently, in the new areas this policy of remuneration has been discontinued.

Immunization programme

The immunization programme has been co-ordinated with the local health centre where they give the vaccines to the health promoters who carry out the immunization campaigns in their own sectors. When the majority of the children are immunized and it becomes difficult to obtain a regular supply from the health centre, those children who require immunizations are referred. Complying with the immunization programme is a requisite of the health programme.

'First-aid box'

Many communities at the introduction of a health programme want to have their own health post, or at least to have a community 'first-aid box'! A number of these boxes were set up, each under the administration of a health promoter. A small charge was made by the health promoter to cover the costs of replenishing the supplies. However, after a few years the demand for this service decreased, partly because the women of the community themselves knew how to deal with many of the health problems that arose, and also because there are now more health services available in the area than when the programme began.

Referrals

The health promoters refer children who require medical care to the local health centres or Lima hospitals. In the case of severe malnutrition or chronic diarrhoea, patients are referred to the out-patients clinic of the Nutrition Research Institute. There are programmes for the follow-up of patients with tuberculosis in the health centres and the local parish. An excellent school for disabled children exists in Pamplona Alta and a religious sister runs a programme to which children with these problems can be referred. The nutritionist has helped in the orientation of the cooks in the menu planning for a *'comedor popular'* (lunch programme) in a local market. These meals are subsidized and needy families can be referred.

Relationship with other health programmes

Co-ordination with the Ministry of Health has existed since the beginning of the programme. Contact was initially made at the central level but the working relationship has been with the personnel of the local health centre who have supported the programme through contributing to the training of the health promoters and in the supply of vaccines.

From the beginning of 1984 the municipal local government has organized campaigns for the prevention and treatment of diarrhoea and for a milk programme for children under six years of age. The health promoters were, and continue to be, active leaders in both these campaigns (Choquehuanca 1985).

One of the sectors where the programme initially began to work 10 years ago had its health 'modules' built by the community and UNICEF, and supervised by the Ministry of Health. Work is principally done by the health promoters of the community. Consequently, the health and nutrition programme has now retired from this sector and is concentrating upon the new '*ampliaciones*' where there are as yet no other health facilities. The women in the sector that was left understood and were supportive of the decision to move on.

Results

Coverage   The programme has served a community that has increased to approximately 4000 people over the years. However, the average number of children followed in the health programme has been about 50 per cent of the estimated under-five population. An increasing number of women work to sustain their families and for this reason are unable to participate. There are a number of 'food for work' programmes in the area, and these do not facilitate the participation in the health programme. Some mothers do not wish to participate as they do not yet see the need for health checks when their children are not ill. The health promoters, however, are aware of the situation of these families.

Nutritional status evaluation   Figures show that apart from an initial decrease in the incidence of malnutrition there has not been very much change over the years. However, the economic situation has deteriorated dramatically, and it may be that the health programme compensates for this to some degree.

The level of malnutrition observed in the programme is greater than that reported by surveys in other poor areas of Lima, e.g. Lopez *et al.*'s study of Huascar (1985). This is probably due to the fact that the programme attracts families with children who have a health or nutrition problem, even though it is publicized as a 'well-child' programme. The new '*ampliaciones*' are more recent, less well established, poorer, and do not have access to the same services as the Huascar area.

It is interesting that the greatest reduction in malnutrition has been in the under-one-year age group. This coincides with the increased incidence of exclusive breast-feeding that was observed in the programme, and possibly improved weaning practices. However, the malnutrition in the older age group has not decreased similarly and may reflect the effect of the critical economic situation on family diet.

Evaluation and discussion

Nutrition and health status   It is difficult to say if the programme has had
an impact on the health and nutritional status of the population, although
the mortality appears to have decreased. However, there are changes in the
health practices that can be observed: a larger attendence for growth monitoring, an increase in duration of breast-feeding and a greater coverage in
immunization. These reflect an increased awareness and demand for health
services, and a greater concern for preventive health care.

Curative versus promotional   In the health programme it was difficult to
maintain a balance between curative care and promotion and prevention.
With the health needs of the population it is easy for the former to
predominate. However, with the change of emphasis in the role of health
promoter as motivator and promoter in her community, promotional and
preventive aspects have been strengthened.

Health promoters and the community   The attitude of the community
towards health promoters has also changed, and they now accept them as
health leaders. This is partly demonstrated by the attendances at the growth-
monitoring meetings which are organized by the health promoters alone.
This has taken time and required constant working with the community,
rather than training health promoters independently of the community and
then expecting the people to accept the alternative health care.

Relationship with the community—women as leaders   The health and nutrition programme has always been invited into a new sector by community
leaders. Initially the co-ordination was with the leaders themselves, but as
the programme developed the work has been predominantly with the women
themselves.

The work with the health promoters and delegates has emphasized their
role as leaders for health in the community, but they have subsequently
assumed other community responsibilities as well. One example of this was
their leadership in a march for water when there was no supply in the community for a prolonged period.

It is through the work of the health promoters together with the community that alternatives can be looked for in improving the health and nutrition
situation of the populations. The health programme cannot solve the problems, but through health and nutrition education a greater awareness of the
problems themselves and their causes can be developed. The health programme contributes both to an improvement of the health and nutritional
status of the children of the '*pueblos jovenes*' and to the community in
working towards alternative solutions.

Comments and conclusions

The Cité Simone, Haiti, programme and the community programme in Lima demonstrate two very different approaches to intervention. In Cité Simone, community involvement and participation, except as a recipient, were minimal, whereas in Lima the programme started at the invitation of the community and the health workers are seen as answerable to their own people rather than as part of the health services. Both seem to have had a significant, if limited, impact and it is tempting to suggest that the ideal would be a combination of the vigorous analytical methodology of Cité Simone with the gentler participatory approach of the Lima project.

The limited value of nutritional rehabilitation units from the point of view of number of beneficiaries, educational impact, and cost effectiveness, is once again demonstrated in the Lima project: the realization that the best place for nutritional education is in the home led to the development of the community programme.

Ways of combating urban malnutrition

There can be no doubt whatsoever that poverty lies at the root of much of the suffering from health and nutrition problems in urban slums throughout the Third World. Yet, as Berg noted, 'Political and economic realities being what they are . . . in most countries it may be a huge task merely to hold the line at existing levels of inequality. Even with moderate success in directing general development into accelerated growth in the income of the lowest deciles . . . combined with a strong effort to increase food production . . . strong complementary measures will be required to increase the level of food consumption of the poor. The expected course of national growth is not a promising means of meeting this generation's shortfalls.' (Berg 1981, p. 24.)

It is clear that 'trickle down' is not the answer for generations to come. Clearly if we are concerned, some sort of income redistribution favouring the poorest people is essential. Every year produces another cohort of infants that are likely to become malnourished while we wait. Even small-scale programmes can have an effect on the nutritional status of children, as shown in the case of a community health programme in Coatzacoalcos, Mexico. Figure 6.7 shows the changing levels of children's nutritional status since the beginning of the programme.

Improvement in the quantity and quality of food   Drawing suggestions from Austin's excellent book (1980) we can summarize possible methods of supplement as:

- food rations/coupon system;
- cheap food shops;
- consumer co-operative shops (making use of bulk purchase);

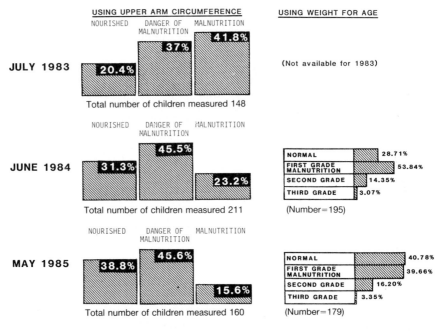

Fig. 6.7 Changing levels of children's nutrition in Coatzacoalcos, Mexico, 1983–5. *Source*: Bosnjak and Cayon (1985).

- food processing;
- targeted food hand-outs to children and pregnant and lactating women, either take-home rations or 'on the spot' feeding such as milk or soup kitchens;
- provision of crèches for working mothers;
- special nutrient foods (e.g.'*faffa*' weaning food in Ethiopia).

More studies are needed in all these areas. In particular, there is very little data on cheap food shops and consumer co-operatives which increase purchasing power without creating dependency.

Even when the distribution of supplementary rations is well organized results can be disappointing: often this is due to a failure to understand families' feeding practices. The take-home rations intended for a malnourished child are likely to be shared round the family or, alternatively, the malnourished child's share of the normal family ration will be reduced. In most situations there is a strong case for regarding the family as a whole as vulnerable and providing a larger supplement of the cheap staple food rather than a small quantity of the relatively expensive supplement. It is, however, hard to generalize: the best solution is likely to be found by discussing the various issues and possibilities with poor families themselves.

When feeding is institutionalized, as in *crèches* for the children of women

helps to design culturally sensitive pgms

building-site workers, it is possible to ensure that the ration reaches the child: however, it is still likely that the quantity of food the child receives on returning home will be reduced.

Food supplements, are not going to be a long-term solution; on the other hand, they are available in many places, much of the time, and have been for decades. They can and should be targeted to the population really in need. The daily milk and soup kitchen programme run by the municipality of Lima, Peru, is an example of this (Choquehuanca 1985).

It is generally believed that hand-outs which increase consumption in the home environment are preferable to institutionalized nutritional units, although the latter do give more control. In Addis Ababa it was found that while the nutritional status improved for children given total feeding in day centres, the overall level of child malnutrition in urban areas deteriorated steadily as the family ration declined due to a lack of cheap food (Goyder 1985). A programme which arrived at a larger preventive ration for the whole family was more effective than the Maternal and Child Health clinics which only provided supplements for mothers of malnourished children to take home.

Legislation Besides establishing ration systems, legislation at central or municipal level can influence the quality of diet; the obvious example is the control of advertising which promotes junk foods, drinks, and artificial milk substitutes. Other possibilities are entitlement to maternity leave, *crèche* facilities, etc.

Education While the impact of nutrition education has often been small, this is usually because the educators do not understand the community that they seek to serve. Often foods or cooking methods are recommended which are too expensive or impractical. Mothers may be exhorted to boil all the water used in the home without taking into account the cost of fuel and time involved. Trying to combine nutritional education with on the spot feeding programmes such as nutritional rehabilitation units and *crèches* often fails because mothers are too busy to learn either because they have employment or other children to care for. The best place for nutritional education is in the home.

Breast-feeding and growth monitoring Perhaps the single most important educational message is the promotion of breast-feeding: its decline is more marked and rapid in urban areas than rural due partly to social and economic pressures as well as to the availability of dangerous alternatives. Once again, sensitive community studies are needed to understand the underlying causes and to plan interventions: a combination of alerting mothers to the advantages of breast-feeding and enhancing the feasibility of their doing so (e.g.

lobbying for breast-feeding facilities in the mothers' working environment) is likely to be more successful than any single approach. Growth monitoring and many of the other elements in the GOBI-FFF strategy (see Ch. 5) will also have a positive effect on the nutrition of infants and young children, whether in slums or rural areas.

Urban gardens   As stated earlier, one of the greatest problems for the urban poor is the need to buy food instead of growing it. While certain areas may be too crowded to allow food production, others may have gardens round their houses or nearby outskirts. In addition, there may be large, open, waste areas suitable for communal or private gardens. In these circumstances, a significant proportion of a population's food can be produced. In Colombo, Sri Lanka health wardens have held nutrition demonstrations with a total attendance of over 3000 by 1981 (see Adamson 1982; Peries 1985). As a by-product, 62 home vegetable gardens had been started by 1982 in the slums where there was space. Similarly, as an extension of a sanitation programme in Orangi, Karachi (see Chapter 9), gardening experts have visited areas that have improved their own sanitation, to give instructions for growing vegetables in the home and to give out free seeds. In nine months there were 1250 'adopters'. Obviously, the quality and extent of cultivation varies a great deal. Some of the gardens, as yet a small percentage, not only supply the household needs, a small surplus is even sold. In other cases it is a marginal activity, a mere hobby. However, the interest is growing steadily (Hasan 1985).

Examples of urban gardens in Rio de Janeiro, and other areas of Brazil, Panama City, and Managua are given by Finquelievich (1985) and Pacey (1978). The promotion of home gardening can appear as a component of health, applied nutrition, or women in development projects. Primary health care projects can include clinic gardens as a demonstration and training centre for home gardening. However, as Brownrigg (1985) has pointed out, gardening components in urban projects to date are rare. The planning of an urban garden should include a local caloric staple (e.g. a cereal) and a legume to supply protein. Mason and Stephens (1981) discuss some organizational options including home gardens with individual plots, individual gardens within a communal area, and communal plots.

Informal food supplies   A considerable proportion of the food in cities is provided by the informal sector, notably the street vendors (there are 300 000 in Lima). These have the double advantage of cheap, labour-intensive processing as well as income generation. Urban householders in Indonesia and the Philippines spend about 25 per cent of their food budget on these street foods. In certain circumstances, small community projects to encourage local processing and marketing will keep money circulating in poor areas: appropriate inputs can include training, credit, and health and hygiene education.

Reduction of infection in children and mothers   This can be achieved by general and specific measures. Typical general measures are improvement of housing, sanitation, water supply, and general health education. Examples of specific measures are:

- The early treatment of infection during pregnancy coupled with the treatment of anaemia.
- Immunization, especially against measles.
- Early treatment of diarrhoea with oral rehydration solutions.

Provision of family planning advice and services   This is particularly important in the case of teenagers who are likely to produce low birth-weight babies. Aggressive family planning policies motivated by demographic rather than health arguments have too often made communities wary of birth-control. However, the World Fertility Survey gives clear evidence of an unmet felt need for contraception in many areas (WFS 1984). As always it is best to start with the community and understand their perception, rather than deciding strategy and objectives centrally. Contrary to the view of some mass motivators, many people are interested in birth-control and want more information as well as sensitive services near their home.

As the experience of an NGO in Bombay has shown, fears and suspicions are best allayed by members of the community using contraceptive methods successfully and then telling friends and neighbours about it (Parikh 1985).

In conclusion

It would seem that for effective widespread results there are measures to be taken at all levels.

Fig. 6.8   Stealing in the city. *Source*: D. Brunner.

- Globally, a more generous attitude towards debt repayment and a food aid policy which is planned to help recipients rather than producers.
- At central government level, legislation providing rationing, advertising codes for foodstuffs, and a health policy emphasizing PHC.
- At city level, special projects such as milk kitchens, *crèches*, provision of family planning services.
- At the community level, health worker schemes, consumer co-operatives, kitchen gardens, lobbying for better conditions, etc.

Finally, 'there is ample evidence that things can be done and we should get on with it'—Wray (1985).

7

Income generation

Poor people seek opportunities for earning rather than
learning.

Income generation and health

There seem to be two main reasons why health programmes become involved
in income-generation projects in urban areas: first to increase the income of
the poor; and secondly to create an income for the self financing of the health
services.

Institutional incomes

The effects of recession, besides increasing the individual's need for an
income, have also led some agencies into seeking ways of increasing the level
of self-funding to ensure survival of the service. Costs have continued to rise
while the community's ability to contribute towards the costs of health
services has often declined. Many health services have found that they are no *costs » Y - poor*
longer able to rely on community contributions to any significant degree.
One solution to this has been the encouragement of income-generating activities by the agencies themselves. It also has to be said that the search for
independent sources of income relates specifically to non-governmental
organisations (NGOs), where they have failed to obtain support from their
own governments and are also wary of excessive dependence on foreign
funding which may be unreliable in the medium term. Most donors are */. self-*
unwilling to enter into long-term commitments to the running costs of *financing*
service programmes. *'h services*
   Operation Friendship in Jamaica has managed to utilize its vocational
training programme to the full by also developing markets for its products
such that it is now able to meet some 40 per cent of the costs of its extensive
training and health programme in Kingston (Brown 1985).
   The Undugu Society of Nairobi, Kenya, introduced an income-generating
programme to provide incomes for voluntary community health workers as
a solution to the problem of trained CHWs 'dropping out' through economic necessity (Ongari and Schroeder 1985).
   It is, however, not our intention to dwell in any detail on agencies' efforts
to self-finance their activities but to look further into the attempts of health
programmes to diversify into activities designed to improve the income of
targeted sectors of their client population.

Increased incomes

There have been two major channels through which health programmes have
approached the need for increasing individual and family incomes. The
birth-control agencies were perhaps the earliest in the field of income-
generating activities, encouraged by the findings of research which found
that the demand for contraceptive methods rose when survival rates for
children improved and general levels of prosperity increased. It was therefore reasoned that to improve the acceptability of birth-control, and ulti-
mately to slow population growth, it was necessary to help families to
increase their incomes in an attempt to achieve overall improvements in
infant mortality figures and a level of prosperity that would reduce the need
for a large family labour force with the marginal income advantage each
member can bring. Whatever the rights or wrongs of this view of income and
health, the attempt by specialist birth-control agencies to venture into
income generation and community development as an indirect way of
reducing population growth has led to a confusion of roles both within these
agencies and on the ground, given that the same personnel have been involved
in both programmes. In practice, the direct link between reducing family size
(or increasing spacing periods between children) and income levels is unlikely
to be measurable or significant in the short term. Also, the rather overly
structured, vertical, top-down approach by birth-control agencies has con-
trasted badly with the more flexible bottom-up approach needed for successful economic development at a community level.

The second link between health and income has emerged through those
health programmes confronting the problems of malnutrition. Conventional health interventions have often failed to have the expected effect on
reducing rates of malnutrition. For example, an otherwise successful programme in Haiti in the Cité Simone area of Port-au-Prince was frustrated in
its attempts to reduce the infant mortality rate (IMR) of severely
malnourished children, despite supplementary and other feeding programmes (see Ch. 6). This led the programme organizers to introduce a
sophisticated form of family targeting, whereby the mothers of the severely
malnourished children were given vocational and other training and helped
to set up small-scale enterprises while the child received nutritional
rehabilitation in the same place. Although longer term education and training programmes were also run for young people, adult men and women, the
specific targeting on mothers of the severely malnourished was an implicit
recognition of the need for increasing the income of the family, rather than
trying just to rehabilitate the affected child. In this way it was anticipated that
there would be a longer term benefit for these children (Boulos 1985).

Similarly, Operation Friendship, the health programme in Kingston mentioned above, identified the children of teenage mothers as being most at risk

from low birth-weight, malnutrition, and early death. The mothers tended to be single, were forced to leave school early due to their pregnancy, and their lack of education made them virtually unemployable, so not only were these young women at a disadvantage in the labour market, but they also had the problem of feeding their children. Operation Friendship thus set up the 'Women's Training Institute' where these girls could complete their formal education and a programme of vocational training, while their children were fed and cared for in a *crèche* on the same premises. Later, these young women were helped where possible to find employment or were encouraged to enter the informal economy using their newly acquired skills in some form of self-employment (Brown 1985).

The necessity to provide basic education as a prerequisite to employment creation has also been identified by a number of groups. This initial activity is often something which can fruitfully be tackled by health groups (see the case study below).

Health activities have proved a particularly useful way of making contact with those women who, for different social and cultural reasons, are often not in touch with development and community activities. In different countries, health acted as this channel, for example in Sudan the Islamic African Relief Agency noted that the use of mosques for health posts enabled women to be able to participate more freely in health-related activities (often the health posts were converted rooms in the mosque, by closing the inner door off and opening a door to the street) (Suliman 1985). Given the increasing percentage of female-headed households to be found in many cities, and the likelihood that these households will have amongst the lowest incomes, any attempts to provide better services and introduce the possibility of increased income have to be welcomed. In Bombay, it was noted that even women living in virtual *purdah* were able through the health programme to increase their incomes through the income-generating and training programme. Again, this was because attendance at a health post was considered more acceptable for such women than other community centres.

It is not only in settled communities that the problem of income inhibits the success of health and other programmes. The Ockenden Venture working with refugees in Port Sudan realized that although the health and other needs of refugees were catered for by the Sudanese government and UNHCR, the key to self-sufficiency for the refugees in an urban area is in employment. Starting in 1982 the Ockenden Venture began to support trainees working with local Sudanese companies. To compensate for the lack of basic skills the Venture also provided some refugees with some basic preparation prior to being placed as apprentices with companies. A handicraft and printing centre for the disabled was also established, which proved commercially successful. For those outside the ambit of these more organized initiatives, a programme of small business loans was established which was of particular benefit to female-headed households. However, this loan scheme ran into initial trouble as people were attempting to set up their activities in those areas already oversubscribed—such as tea shops, delivery services, and the like (OXFAM 1986).

Case study 1 Streehitakarini, Bombay

Based on a paper by Parikh (1985*a*).

Streehitakarini, an NGO in a mainly Hindu community in Bombay, is an example of a health programme which found that to have a successful income-generating programme for women it was necessary first to provide basic education, using non-formal techniques, to equip the women with basic literacy and numeracy skills necessary to enable them to enter into small-scale employment activities (Parikh 1985*a*).

The felt need of the community when Streehitakarini began work with a community of 100 000 was a clinic. The NGO used this as an initial focus for their activities which concentrated on health, education, and income generation for women. By 1972 the programme had 24 voluntary health workers, and now the training of community health workers is a routine activity—with 63 CHWs active in 1986. CHWs not only give lessons on health, nutrition, and family planning, but also take classes for adult literacy and help women save for their future by participating in 'Small Savings Campaigns'—collecting and depositing money with their respective group of women.

Parikh (1985*a*) says of the programme: 'We cannot yet say that the programme is *of* the community. It is *for* the community, practically executed *by* the members of the community, yet it is hard to know how we can make it *of* the community.'

Sewing classes are amongst the most well-established activities of Streehitakarini. In early days women wanted to learn how to sew in order to save money by producing their own clothes. The programme helped women purchase their own sewing and knitting machines. This provision of credit is regarded as very important. Credit is often required to assist groups to enter into small-scale production once they have received the necessary training. Besides helping with the finance required to set up small enterprises, a savings programme was introduced in 1978 based on a State government plan, and by 1982 over 2000 savings accounts had been opened. This sort of credit programme tends to be more successful and easier to manage for the small family enterprise than the larger firm. Credit which is repaid reasonably quickly also avoids the problem of being eroded in value in those countries suffering from high rates of inflation.

One problem identified in many income-generating programmes related to health, is that it does not always follow that health personnel make the best businessmen. If business training is to be given, then it should be good. There have been too many programmes started for the best of reasons, where poor people have been tempted into a small production or marketing enterprise which has been badly thought out and hence lacked viability. A common error has been to encourage people to enter into overly competitive activities, such as sewing which in turn has entailed an even greater expenditure of time and effort for marginal returns. Proper market surveys, economic feasibility studies, or employment surveys should be carried out before people are encouraged to enter into training or financial commitments. In general, it seems that small enterprises with about six people have a better chance of success than larger, more unwieldy enterprises. The smaller unit avoids problems of larger scale and the need for specialized managers, increased overheads, and recurrent costs which may be difficult to meet in a competitive and unstable market. The smaller enterprise will be more flexible in economic terms, and be able to reflect market changes and less likely to develop a dependence on the health programme for managerial and economic support.

Case study 2 Working Women's Forum in Madras

Based on a paper by the Family Planning Foundation, India (1984).
In contrast to the above case study, which Parikh acknowledges is a programme 'for the community', the Working Women's Forum of Madras is an attempt to develop a mass-base that can help women to unite and fight for a better deal themselves. This is significantly different from the elite welfare-oriented groups. This case study is also different from the previous one in that it sets out to be an income-generating programme and then develops into health activities, for example, family planning.

empower? to fight for polt rts

Madras, once known as the garden city of India, has expanded so rapidly in the last two decades that there are now more than 1500 slum areas within the city area. One in four persons residing in Madras is a slum-dweller.

The concept of the Working Women's Forum's unique format and role evolved from the frustration of several women political party workers familiar with the slums, from where they mobilized women for political rallies and forums. They noted the women's low economic and social position and the utter lack of benefits from government programmes—a plight often compounded by confrontation with a system which ignored the civic needs of a population that was either encroaching on, or simply overcrowding, the existing facilities. It was against this back-drop that a long-time political worker—Jaya Arunachalam—decided to quit the political arena and to work towards a non-political alternative that would attempt to give the nearly half a million slum-dwelling women of Madras a platform from which to battle for their rights. Thus in April 1978, the Working Women's Forum was established (Chen 1983; Family Planning Foundation of India 1984).

The Working Women's Forum consists of small neighbourhood groups of 20–25 women coming together and electing a group leader from amongst themselves. The group leader automatically becomes a member of the Working Women's Forum's (WWF) Governing Board—which meets regularly to take all decisions, so that there is direct participatory management of the activities of the organization by its members. Arranging access to credit has been the chief instrument with which the WWF struck its roots. WWF's first efforts were to help women engaged in petty trade, cottage and household industries to acquire easier credit and more remunerative prices. But it soon moved on to help women tackle a number of other social problems. Beginning with an initial membership of 800, the organization rapidly grew to 6000 within two years.

Grass-root support for family planning

The Family Planning Foundation (FPF), an Indian NGO promoting and funding innovative programmes judged capable of replication, joined with WWF to develop a new model, hoping to attract other women's organizations to adopt a primary role in the family planning cause. Joining together, the two groups shaped The Experiment in Leadership Training, a three-year action demonstration project to develop grass- root workers.

A methodology was developed which enabled WWF to take up this work of promoting leadership among women with special emphasis upon primary health care and family planning. Most specifically: (1) a cadre of 60 women workers were to be trained to be community leaders and assist in the conscientization of women's status issues; and (2) communication materials were to be developed, basic data collected, and the entire experiment carefully documented with a view to eventual replication.

Training

Following six weeks of intensive involvement (two weeks spent on basic sensitization on health and family planning, two weeks carrying out the census, and two weeks developing an appropriate action plan based on these needs), the project was formally launched on 2 October 1980, the anniversary of Mahatma Ghandi's birth.

Initially there were difficulties—hostility as the workers tried to elicit information, and animosity from paramedicals as the women ventured into areas which had previously been within their domain. Through patient, empathetic relationships, often to the extent of providing household assistance at times of severe need, the programme gained acceptance. Though the initial response to family planning was more difficult to establish than other health needs, eventually the image of these worker's leaders as women who personally knew and evaluated the benefits of family planning added to the

credibility of family planning. Linked to the economic and general development effort which the Working Women's Forum provided, the primary
health care and family planning work acquired a holistic and multi-
dimensional image never accorded it before. The substantive impact of the
programme, in immunization services, health and nutrition education, assistance in natal care, and family planning acceptance, was so apparent that
even before the end of the three-year experimental period a government
grant became available to WWF to further extend its activities across all the
slum areas of Madras.

Policy implications

This experiment demonstrates that poverty need not be a barrier to realizing
the importance of health and family planning measures; that poor, barely
literate women can be the agents for this work; and that it is a cost-effective
model, capable of replication in other city slums or rural areas wherever a
strong women's network exists.

ind productivity

h ⇌ ec devt

Comments and conclusions

An underlying assumption behind many development programmes, especially those which are service-based, is that improved national incomes will
allow the population to be able increasingly to afford and contribute to the
upkeep of services such as health. Although it is possible to argue that in the
long-term improved health implies greater individual productivity, it is
unlikely that this will manifest itself in the short term. The failure of productivity to increase dramatically as a result of expenditure on health and other
services is likely to be even more marked in urban areas, where the marginal
productivity of labour is likely to be very low due to the high proportion of
the labour force being involved in services, petty trading, and domestic
work. The labour input in agriculture, especially at peak times during the
agricultural cycle (such as harvesting), is often far more likely to be responsive

Fig. 7.1   The surprises of the city. *Source*: D. Brunner.

to an improvement in health. However, the failure of malnutrition rates to fall, despite improved health services, has been but one indicator that real incomes of individuals and nation states have not risen but have often actually fallen. It is no longer possible to assume that wealth will increase sufficiently to permit either the state or the individual to meet the costs of service delivery. At a local level it is the recognition of the longer term nature of economic recession and a more realistic appraisal of the importance of the 'informal economy' to the survival of the poor, that has led health programmes into reconsidering their involvement in actually seeking ways of alleviating the poverty which so forcefully contributes to malnutrition and ill health.

$\uparrow Y < \uparrow P_{food}$

maln still up

8

Health education and communication

Promotional and educational approaches

Health education has become a broad subject and is now closely linked to health information and communications, with a focus on the community level rather than the individual person. There are two main approaches which have emerged recently that incorporate these changes. These are the 'promotional' approach that aims at changing specific health-related behaviour, largely through social marketing and the use of mass media, and the 'educational' approach aimed at showing people how to analyse and change their own conditions. This chapter is mainly concerned with the latter approach.

A useful definition is as follows: 'Health education in primary health care aims to foster activities that encourage people to:

- want to be healthy
- know how to stay healthy
- do what they can individually and collectively to maintain health
- seek help when needed.' (WHO 1983.)

Current concepts in education, information, and communications for health have a long history, and it is useful to trace their antecedents in order to understand how and why they have changed (Walt 1984; Walt and Constantinides 1984). Two separate trends are apparent. The first emerged from the health sector itself, and addressed health problems through health education in an expert-directed 'top-down' approach. The second emerged *promo'l* from community development work, and focused on health care as only one of a number of efforts defined by people to improve their own community *educ'l* well-being, a 'bottom-up' approach.

In its earliest days, health education mirrored what Mahler has characterized as WHO's paternalistic attitude to member states, where the WHO 'Decided what was good for them, and set up WHO projects to prove how right it was.' (Mahler 1982.) Health educators believed that if people would only listen to, and act upon, their advice, they would enjoy better health. For years the KAP model—giving Knowledge to influence Attitudes and so change Practices—was the *sine qua non* of health education. Methods tended towards didactic teaching of 'hygiene' in schools and in health clinics, and was typically a one-way downward information process. Content related to the preventive and curative aspects of specific diseases and the improved utilization of health services.

Community development, on the other hand, started as a mass education activity for the rural poor. From the outset it was concerned with working with communities to change behaviour by imparting new skills and knowledge. Although health was part of some integrated rural projects, it was not normally a focus for community development. Communications support for such development projects was aimed at accelerating the implementation of projects and at making them more effective. However, there was often a gap between community priorities, and the 'product' on offer (Bunnag 1981), and participation was often relegated to a contribution by the community of money, time, or labour. Informed concurrence and involvement in the development processes from the early stages was rare.

By the 1970s dissatisfaction with the role of health education and communications in development projects was paralleled by many other changing ideas regarding health. The costs of health care were rising and yet still

Fig. 8.1   Health education posters in Kebele 41, Addis Ababa, emphasize the use of latrines, bins for rubbish, and cleanliness in the home. Photograph by *Redd Barna*.

services were not even reaching poor populations. Disillusionment with medical science and technology and the marginalization of traditional and lay knowledge were compounded by the acknowledged limitations in the ability of health services to improve people's health. There was a call to improve basic services to improve people's lives, to' . . . expand and improve facilities for education, health, nutrition, housing and social welfare, and to safeguard the environment.' (United Nations 1970.)

These and other influences converged in the Alma Ata Conference on Primary Health Care in 1978, and became part of the broad policy thrust of Health for All by the Year 2000. As noted in Chapter 5, the first of the eight essential activities for PHC was listed as 'education concerning prevailing health problems and methods of preventing and controlling them' (WHO 1978), and later health education was given an explicit role. 'Its primary responsibility is to promote individual and social awareness leading to people's involvement and self-reliance.' (Mahler 1982.)

Thus, in the 1980s new approaches to health have highlighted the need for both the bottom-up and top-down approaches. There is a recognized need to create community awareness, so that communities can themselves contribute to and take responsibility for health improvements, as well as having experts and national policy-makers provide advice, support, legislation, and resources to health priorities. The present focus of health education is to develop the process whereby both approaches can address the same sets of health problems.

In analysing the educational and community approaches in health education in primary health care, there are four main phases: analysis by the community of its own situation; choices of what action should be taken; mobilization of community and other resources; and a consolidation of the community involvement activities and organizations.

The following case studies illustrate these phases well. The Casa Amarela project in Recife, Brazil, is in the early phases and looks at what people can do for their own health in a poor shantytown, emphasizing preventive health education and traditional remedies for ill health. The community-based health programme of the Undugu Society of Kenya illustrates how communities can collaborate with outside agencies to establish joint activities. The Urban Community Development Project started in 1967 in Hyderabad, India, and focused on a wide range of activities, including income generation, housing, and education. This case study illustrates well how joint activities can be consolidated and scaled up so that they become well established and promote long-term urban development.

Case study 1   Community health education in Casa Amarela, Recife, Brazil

Based on a paper by Carriconde (1985).

Living conditions

Poor health and chronic sickness of one sort or another are facts of life for the thousands of people who crowd into the shanty suburb of Casa Amarela on the hilly northern outskirts of Recife. They are part of the continuous flow of migrants from the arid interior of Brazil's north-eastern states.

Casa Amarela is an area of high unemployment, where less than half the working population have stable jobs with social security benefits—and even these jobs are poorly paid. Most make some sort of living from a range of odd jobs—car washing, selling lottery tickets, shoe-shining, laundry—and even the children must become bread-winners, such as water-carriers. Few homes have piped water.

The rainy season brings hazards. The steep, unpaved streets become rivers of mud, making higher areas almost inaccessible. Severe erosion causes houses to collapse, killing and injuring some and leaving others homeless.

A local woman reflected on the effects of poverty and desperate living conditions: 'In more than twenty houses on my street where six to eight people live in each house, only one person is working—the commonest illness is hunger and the psychiatric problems caused by hunger'.

Amongst adults, tuberculosis and mental illness are common. Children suffer from parasites, respiratory diseases, gastroenteritis, polio, measles, diptheria, and chickenpox. Social problems lie at the root of most of the ill health in the area—problems which will not be solved overnight. Health education, simple traditional remedies, and community action can however, make considerable inroads.

Community health education

Based on the premise that 70 per cent of illnesses can be self-treated, in contrast with the myth created by the consumer society that health comes only through doctors and their prescriptions, a small team in Casa Amarela has developed a programme which focuses on health education and the revival and promotion of traditional herbal remedies.

In 1980, several communities in the area were already campaigning actively for their rights to the land they occupied, and these associations formed the nucleus for the health programme, covering five districts and a population of over 150 000.

Health workers have promoted mothers' clubs and womens' groups who discuss ante- and post-natal care, family planning and general hygiene and nutrition, and this is followed up with house-to-house visiting. Street meetings, stimulated by the team's audio-visual presentations, discuss local health issues and analyse general problems of the area in a wider social context. 'Clean up' campaigns are instigated, clearing and burning refuse. Various groups have started petitioning the authorities for adequate piped water supplies and regular rubbish collection. Regular four-month training courses ensure that local voluntary health workers can carry on and extend the work so that the health programme becomes an integral part of each community's responsibilities. There are already two or three 'graduates' of the courses working in each area, helping out at clinics and with promoting health discussion groups.

Popular medicine

Against this background of increasing health awareness, the programme is also encouraging people to look towards more traditional remedies for many of their commoner ailments. Although drugs are obviously important in medical treatment, the project abhors their excessive and often unnecessary use by poor people.

In an area where formal health facilities are few and far between, most people rely on pharmacies to provide their medicines. They buy expensive and often inappropriate drugs—or even useless 'tonics'—over the counter, for illnesses which clean water and adequate diet might prevent, or which herbal remedies could alleviate. Yet Brazil has a rich reserve of 'popular' medicine—knowledge once common in rural communities about the healing or soothing properties of numerous plants (for instance the tranquillizing effects of lemon-grass tea, or an infusion of passion-fruit leaves; garlic to control some internal parasites; parsley as a diuretic).

Many of Casa Amarela's older residents remember the traditional medicines and are encouraged to share their knowledge and grow the plants and herbs around their homes, often becoming the local advisers to whom the clinics refer patients. Herbal 'dispensaries' are springing up within the community as local health workers also develop herb gardens and patients are becoming increasingly confident about these forms of treatment. Tapping and using this knowledge is one of the health programme's most significant features. Apart from herbal remedies being cheaper, more accessible and often more effective, their use reinforces people's belief in their traditional values.

The programme as a whole works to strengthen community bonds among a fragmented population who have come from many different states and areas, and to convince them of their ability to control significant factors affecting their own health. The constant exchange of knowledge and experience which the herbal 'pharmacies' involve plays an important part in this process.

Case study 2   Undugu Society, Mathare Valley, Kenya

Based on a paper by Ongari and Schroeder (1985).

Poverty is enormous in some of the communities where Undugu works, and the support offered by traditional rural society simply does not operate in the city. Many people have become destitute. However, Undugu does not believe in taking away the community's responsibilities. It prefers to assist the community to look after the poor.

The Undugu Society of Kenya ('Undugu' means brotherhood or solidarity) is an organization and a movement based in Nairobi which grew out of the work of Father Arnold Grol. In 1973, in a parish where the majority of the parishioners were slum-dwellers, he organized some youth groups. Coming together for such activities as music, soccer, and volleyball was attractive to young people and it also gave them the opportunity to discuss how to survive in a situation where so many despaired. They asked for training. Father Grol started a small carpentry workshop which grew into a technical school. By 1985 this village polytechnic in Mathare Valley, fully recognized by the government, functioned under a local committee independent of Undugu itself.

Father Grol made contact with the street boys of Nairobi—the so-called 'parking boys'—and eventually a house for them was opened in Mathare Valley, together with four schools.

Contact with the parking boys led to contacts with their families, often one-parent families. Undugu assisted initially by sponsoring some of the children in primary schools, and they also discussed ways for families to make a better living. This was the beginning of many women's groups. The women began by knitting pullovers for school uniforms and making crochet dresses and other handicrafts. But once they had formed a group, they then began to discuss their own problems. New ideas developed, such as the health programme in which the women of the slums are trained as voluntary community health workers.

All Undugu's programmes have been funded by donations from friends and organizations. The Kenya Government, for instance, partly pays for the salaries of some of the teachers of basic education. But just as Undugu does not want to make people in the slums dependent on Undugu, so, too, it would not be right if Undugu itself remained fully dependent on outside-aid agencies. That is why Undugu has started income-generating projects where training and making a profit go hand-in-hand.

What began as a youth group over 10 years ago has now grown into an organization for internal community development. Undugu now employs more than 75 people and provides some income for hundreds of people in the slums. 'We know that we cannot provide all the answers, but we want to encourage people to continue the search' (Ongari and Schroeder 1985).

Fig. 8.2 Children of Mathare Valley, Nairobi, where the Undugu Society programme is located. Photograph by Hartley, OXFAM.

In October 1980, a one-day workshop took place at which experts in community health from different organizations came together to discuss: 'Is the concept of the rural community health worker applicable to the urban environment?' It was decided to experiment with a community health programme in the slums of Nairobi. Undugu got a lot of support and assistance in this attempt from the Community Health Worker Support Unit at AMREF (African Medical Research Foundation) in Nairobi.

The Undugu community health programme

Following the Undugu philosophy, the health programme is a self-help project. The main objectives are:

- raising awareness in the communities in respect of their responsibilities in the field of preventive and promotive health;
- training voluntary community health workers (CHWs) in low-income settlements to assist in preventing disease and promoting health care, with special emphasis on women in the childbearing age group and children under 5 years old.

The community-based health care programme is the part of health care which is literally based in the community. The whole community needs to take part in it, not only a few volunteers trained as CHWs. Community-based health care (CBHC) is like an African woven basket *'kiondo'* (see Fig 8.3). The basket is made up from many strands and the weaving starts with such people as community elders, church leaders, women group leaders, youth groups, extension workers, trainers, CHWs, health committee members, and traditional healers. The long strand, which then connects all the other strands, symbolizes the people themselves, co-operating with all the above-mentioned leaders and groups coming from within the same community. Unless the health programme is carried out by the community as a whole, it cannot last. What is the basket without the one strand or the other? It will not be useful and beautiful and durable. The same applies to the health programme.

In order to facilitate the start of an effective community-based health care programme in a community, a wide and intensive conscious awareness-raising has to take place. People in the community have to realize that there is a need for preventive health measures and that they have two possible approaches, communal and individual. They have to feel determined and committed to do something themselves for this programme in their own community. If the people say 'Yes, we want community health workers' that is no base for a programme, it is a promising start. The people need to say 'We ourselves will be responsible for this programme.'

The health committee

An early priority is the formation of a health committee elected at a public meeting with a wide representation of the community.

It is the task of the health committee to:

- help the community to elect community health workers;
- give moral support to the community health workers;
- co-operate with them in deciding what are the major health needs of the community, setting goals, planning action and evaluating efforts;
- motivate the residents for community actions towards better health;

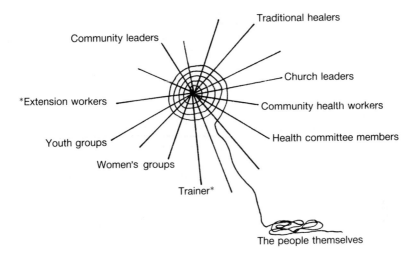

Traditional healers

Community leaders

Church leaders

*Extension workers

Community health workers

Youth groups

Health committee members

Women's groups

Trainer*

The people themselves

*Not technically community based

Complete basket (kiondo)

Fig. 8.3 Community-based health care can be compared to the African woven basket. The people themselves are connected to groups in the community. What is the basket without one strand or the other? *Source*: Ongari and Schroeder (1985).

- resolve conflicts between community health workers and residents;
- be a link between official health facilities and community health workers.

Members were trained for three weeks, learning about co-operation and leadership skills, their role and relationship to the community health workers.

Community health workers (CHWs)

The CHW needs to be a motivator, a communicator, and a good listener in order to know what is going on in the community. They should be able to communicate effectively and motivate residents to change their habits and attitudes towards better health. They are local residents and elected by local people.

The CHW starts by visiting homes around her own house, advising people on preventive health measures and home remedies. The CHWs do not have any medicines, because the health facilities are close to the people. Undugu also believes that CHWs with drugs would soon put too much emphasis on the curative aspects of health. The CHW visits 20–50 homes, with a maximum of 100. Besides home-visits, the CHW uses public village meetings, group meetings, school visits, etc. in order to motivate the community.

Training for CHWs puts an emphasis on communication skills to enable them to advise on prevention, early diagnosis and, when possible, to use low-cost home treatments (e.g. oral rehydration solution (ORS)), and to make people aware of the available medical services. The flexible curriculum covers environmental and personal hygiene; diet, breast-feeding, and malnutrition; pregnancy, delivery, child care, and family planning; and common illnesses and their prevention. The training is carried out over a long period, with theory followed by several weeks of practice.

Experiences of the Undugu community health programme
The health programme started as a pilot project in Kitui Village, Nairobi, because this was a well-organized community with an active village committee and a limited population.

A community survey was done to learn about local health problems which showed that the important ones were malnutrition, diseases caused by lack of personal and evironmental hygiene, and respiratory tract infections, particularly tuberculosis. The Undugu health team then discussed these findings with the residents.

In February 1981 the village committee selected 12 women who volunteered to be trained as community health workers. The basic training took place daily over 9 weeks.

The Undugu health team and the CHWs drew a map of Kitui village and each CHW took over the responsibility for a certain part of the village—starting around her own home—with 40–50 families on average. Each CHW —even the illiterate ones—received a home-visiting book. In 1983 the CHWs started working closely together with the family-health field educators from the Ministry of Health. The CHWs were well accepted but the community did not give them much support.

A major problem was the severe lack of an income for the CHWs. Undugu did not want to create dependency by paying CHWs, but with an empty stomach it is difficult to learn or volunteer to help and motivate one's neighbours. Therefore, in 1984, discussion about income-generating activities began. CHWs decided to contribute some money from their own pockets (100 Kshs, = US$6.25) and asked Undugu to contribute as well so that they would be able to start a tea kiosk. But it took so long for the money to come that only now are they establishing their kiosk.

In August 1981, Undugu staff started to get acquainted with a neighbouring settlement called Kinyago. Home visiting started in 1981/82. At the same time, public health lessons were held once a month in the village, and the concept of community-based health care was thoroughly explained. A few young women volunteered to undergo the training, but there was no backing and support from the community. Undugu started training these women, but already in the first sessions it was clear that there was not enough motivation. Undugu sat down with the village committee, and it was decided to stop the training and to wait until there were really motivated and dedicated candidates from the village. At the same time there were other organizations in the village, who were working in the opposite direction, by offering curative treatment in the village and by transporting sick people to Kenyatta Hospital in Nairobi. The Undugu health team decided to leave in 1983, saying that they would return whenever the community was ready. In 1984 and 1985 meetings were called but the problem of support was unchanged. Since the community does not seem to be interested, it is clear to Undugu that there is no way to initiate a real community-based health care programme.

The Undugu health programme expanded at the beginning of 1983 with home visiting in 'Mathare Valley village III' which is said to have approximately 20 000 inhabitants. A community sense hardly existed there compared to Kitui village. Therefore Undugu decided to work through a prayer group. A meeting explained the health programme and the group members were visited in their homes to find out about the health problems in this area. The high number of physically and mentally handicapped children was striking. The need for CHWs was seen and in May 1983 the group selected 32 women from amongst themselves to be trained.

The basic training in Mathare took much longer than in Kitui, and after an attempted *coup d'état* in August many women went home to the rural areas and only 19 completed the basic training. In late 1983, a public village meeting took place in which the Chief handed over the certificates to the CHWs. The CHWs received a teaching-aid kit to be used in the village for educating mothers on health.

The Mathare CHWs also lacked an income but were unable to agree how to earn one collectively. Another problem in Mathare is that the population turnover is very high and the number of CHWs remaining by 1985 was only seven.

The total number of slum dwellers in Nairobi is close to half a million and Undugu does not expect to reach even half of them. Undugu sees their health programmes as a model for community-based health care, which in time should be organized jointly by the public health services of the city of Nairobi and the local communities themselves.

Lessons from Undugu

Ongari and Schroeder summarize the lessons from these programmes:

1. Get to know all the agencies working in the area and find out what they do.
2. Have good inter-agency communications and be generous with feed-back.
3. Try to encourage all the agencies to follow one policy which encourages people to be self-reliant.
4. Before starting community-based health care, intensive overall awareness-raising is necessary so that the whole community can be involved.
5. Start in an area where the entire community wants the programme.
6. Start the programme with the health committee, not with the community health workers.
7. The CHWs should be elected by the entire community in a public village meeting and not only by the village elders.
8. Start an income-generating project for the CHWs as soon as possible to compensate for their lack of income.
9. Set clear objectives—be clear about what is to be achieved.
10. Easily implemented health projects should be introduced first with CHWs involved in project planning and implementation.
11. In the CHW basic training, do not include the months of the school-holidays or seasonal migration.
12. Do not expect people in a learning situation to attend a long period of theoretical training lasting weeks or months; organize it day by day.
13. Do not give foodstuffs for income compensation during training because it creates dependency.
14. Motivate trainees by giving them recognition through involving government officials and leaders, the chiefs, and the district officers.
15. Teach CHWs communication skills, so that they are able to share their knowledge effectively.
16. Train CHWs and health committee members according to their own pace, because they are the ones doing the learning.
17. Encourage CHWs to be very strict about home visiting in their own local community, where everyone knows each other.
18. Ask CHWs to collect only data they understand and which they can use in their daily work.

Case study 3 Urban community development project, Hyderabad

Based on a paper by Rau (1985).

Introduction

Urban community development in India had its origins in the large-scale rural programmes which were adapted for urban pilot projects in Delhi, Ahmedabad, Baroda, and Calcutta. The National Planning Commission then reviewed this experience and recommended a further series of projects which were to be initiated by the Ministries of Health and Urban Development.

Hyderabad is a large, rapidly growing city in central India with a Municipal Corporation that has been responsible for urban development and which undertook to support the Urban Community Development (UCD) project. In fact, the UCD project is an organizational part of the Hyderabad Municipal Corporation.

The aim was to use local resources through enhanced popular participation in local community affairs, self-help, mutual aid, and increased co-ordination between government and non-government agencies, particularly once local community needs and priorities had been identified.

Self-help is the central motto of the Urban Community Development project, which started with the following assumptions:

- Every person should be involved in making choices and in the activities.
- Community development starts where the people are and helps them go to where they want to go.
- Any community, however poor it may be, will be able to do something to improve its living conditions through its own resources, leadership and organization. Outside help should only be sought after fully exploring and utilizing local resources.

Urban communities and self-help

The objectives of the programme include:

- Creating a sense of social coherence on a neighbourhood basis through encouraging corporate civic action and promoting a sense of local integration.
- Developing a sense of belonging to the urban community through increased participation in community affairs and enabling people to solve problems using their own initiatives, organizations, self-help, and mutual aid.
- Bringing about a change in attitudes by creating civic consciousness and by motivating people to improve their own conditions, particularly those affecting their social and physical environment.
- Developing local initiatives and identifying and training local leaders.
- Ensuring the full co-operation of the technical and welfare services offered by the Municipality and other organizations.

Components of the project

The project started on the basis of felt needs with every programme being identified by the people. The role of the UCD is to help in such identification, keeping in mind the broad objectives. UCD programmes need to be popular and inexpensive, as well as flexible and replicable, since they are based on people's initiatives, expressed needs, participation, and local resources. For all these programmes human motivation is the prime consideration and money is secondary. There is no free distribution of money or other inputs, and no programmes are initiated unless and until there is sufficient response from the people. Only activities which can be taken over by local agencies are initiated in the first place.

The response has been very rapid. The Hyderabad Urban Community Development started in 1967 with one pilot project covering a population of 50 000 and has grown into nine projects covering the entire population of the twin cities, with a great emphasis on the slums. UNICEF started assistance in 1977, sharing equally the support to six projects with the State Government. The support to the remaining three projects is borne entirely by the Municipal Corporation and the State Government.

A lot of awakening has taken place in slum areas, as a result of which there are now 160 *Balwadies* (pre-primary schools), 98 *Anganwadies* (child-health clinics), nine *crèches*, four coaching centres, 101 adult literacy centres, 17 reading rooms, 15 music and dance centres, four women's typewriting and shorthand centres, 98 sewing centres, and 35 integrated health-care centres in the slums.

The Urban Community Development has also established 99 *Mahila Mandals* (women's organizations), 135 Youth Organizations, 223 Basti Development Committees, and arranged economic support loans to 2762 families during the last 3 years.

The emphasis in community health programmes has been on mothers and children. A special initiative for improved immunization coverage against the major communicable diseases of childhood is being carried out. Local community organizations participate by providing the services of community volunteers who help the medical teams conduct surveys of children aged 0–6 years. These volunteers have also co-operated in arranging publicity campaigns about the immunization programme.

Women leaders and representatives of women's organizations have been trained to become trainers. These have in turn conducted over 90 one-day training camps for over 3000 slum housewives, which have covered the following topics: oral rehydration, breast-feeding, female literacy, child care, nutrition, health education, and family spacing. The UCD project, working with women who have poetic and musical talents, has developed publicity songs which are distributed on tape cassettes.

Nutrition Centres are run in 212 slums and are under the management and supervision of voluntary organizations. About 40 000 children up to six years of age, and pregnant and nursing mothers are covered under these programmes. Growth charts are introduced and mothers are trained how to use them for their own children.

Medical check-ups and eye camps are conducted periodically, with the co-operation of service organizations and local medical colleges in which doctors offer their services free to the urban poor.

Low-cost sanitation

Special emphasis is given to low-cost sanitation in slum areas where public services are very poor. Households are encouraged by incentives to construct their own individual latrines and to convert dry-earth latrines into the water type. In all the housing colonies which are constructed by the people on a self-help basis, every house is provided with an individual latrine. Sewer lines are laid up to the doorstep of the slum houses and the dwellers have readily constructed their own water closets.

Periodic sanitation drives are conducted by the community—for cleaning drains, levelling the ground, building approach roads, constructing community halls, and improving the pavements. This all encourages keeping the neighbourhoods clean. The latest approach has been preparation of comprehensive mini-plans by neighbourhood committees for basic services at the slum level. This approach involves the people at the grass-roots' level, who thus become more aware and plan and implement their own activities. The neighbourhood plans include programmes for health, nutrition, education, as well as water supply, sanitation, income, and housing.

Housing improvements

In 1979 there were 455 slums in Hyderabad, with nearly half a million people. Of these slums, 137 were on government land and the rest on private land. Most of these slums have existed for more than 20 years and some of them are over 40 years old. Of the 137 slums on government land, 20 have been declared unacceptable in view of their location according to the City Master Plan. The remaining slums are being regularized by giving tenure to residents and by improving the environmental conditions.

The Municipal Corporation of Hyderabad, under the auspices of the Urban Community Development Programme, has a housing programme on a self-help basis, and during the period 1975–81 nearly 3000 houses were constructed in 31 slums, with loans from commercial banks. The loan per house was for Rs. 4000, out of the estimated total cost of Rs. 6000 for each house. The Corporation agreed to the construction of another 6000 houses in 61 slums during 1981–5. Each house cost about Rs. 9000, of which the loan assistance was Rs. 6000, the government subsidy Rs. 1000, and the balance of Rs. 2000 was contributed by the householder.

Slum improvement

The Municipal Corporation, following on the success of the UCD project, began the systematic development of slums under the Master Plan Improvement in 1981–2, with 228 slums selected for provision of basic amenities such as roads, sewers, storm-water drains, public latrines, water supplies, and street lighting.

The Corporation has extended these improvement activities in the second Master Plan to a further 207 slums, with assistance from the Overseas Development Administration of the UK Government. This programme was launched in 1983–4 and will be completed by 1988.

Out of the several earlier UCD pilot projects 'the only successful survivor is Hyderabad, which at present covers 80 percent of the slum population of the city. It is held out as a model for other states.' In Andhra Pradesh itself, similar projects have now been started in Visakhapatanam and another at Vijayawada.

Comments and conclusions

From these case studies several important conclusions can be made:

- They demonstrate that the community development approach to the problems of the urban poor can be effective, both in providing improved social and health services cheaply and also in meeting many of the basic problems of poverty.
- They also show that despite poverty, there are potentially large economic resources in most slum communities which can be realized with the help of sensitive community development workers.
- They show that a major key to success in community development is the project staff and their approach. In fact, the selection and training of staff is probably the most important element in any community development programme.
- They demonstrate the importance of co-ordination in achieving the most effective use of community, local government, and external resources.
- They also demonstrate the importance of certain basic linkages in the development process. These include (i) integration of physical improvements within the community development process; (ii) systematic linking of voluntary organizations with slum communities; and (iii) systematic linking of slum residents with financial institutions in the formal sector of the urban economy.

Fig. 8.4   Education in the streets. *Source*: D. Brunner.

9

Sanitation

Problems of sanitation in the poor urban environment

For the poor, much of the health risk in urban living arises from poor sanitation. Cairncross (1985) has produced a comprehensive review of 'sanitation and the urban poor' and this chapter draws largely upon that work.

High population densities of urban areas often render rural forms of sanitation technology useless or even dangerous. Pit latrines fill up, leaving no space available to dig a new one, and wells become polluted by nearby septic tanks. In many urban areas, only sophisticated and capital-intensive forms of infrastructure, such as water mains and sewerage networks, seem to offer a chance of providing an adequate standard of public hygiene.

Providing even minimal sanitary infrastructure is particularly difficult in the cities of developing countries, for two main reasons.

First, the cities are growing extremely rapidly (as described in Chapter 2).

Fig. 9.1 Homes on stilts in Recife, Brazil. The location of many poor urban areas makes environmental hygiene very difficult. Photograph by Pearson, OXFAM.

Secondly, they have often developed in quite unsuitable locations. This is particularly true of the port cities developed for trade with the colonial powers; Lagos, for example, or Bombay, which have grown up on coastal marshes subject to flooding, infested by mosquitoes, and in which any excavations, for pit latrines or pipe trenches, rapidly fill with water. The inland cities like La Paz or Addis Ababa are often in hilly or mountainous locations, but the unpaved streets suffer erosion in the tropical rainstorms, exposing services and often causing landslides or the collapse of houses.

Five major areas of sanitation for the urban poor can be considered:

- water supply
- rainwater drainage
- solid waste disposal
- 'sullage' or grey water drainage
- excreta disposal

'Sanitation', in its broadest sense, can apply to other activities, particularly the control of disease vectors such as mosquitoes, snails, and rats. Improvement of sanitation in the five areas listed above will help to control vectors, but vector control as such is not considered here. Nevertheless, urban environments provide particularly good opportunities for vector control; the breeding foci are few in relation to the numbers of people who can benefit by their elimination, and municipal staff can work on vector control during periods, such as the dry season, when their load of repair and maintenance work is relatively light.

Water supply

The urban poor frequently pay an exorbitant price for water. White *et al* (1972) found that those buying water from kiosks in Nairobi's Mathare Valley were paying 10 times as much per litre as the wealthier citizens with private connections. The concessions to operate these public water kiosks are mainly in the hands of prominent political figures. Lusk (1982) found that a 20-litre tin of water, sold by an itinerant vendor, cost a quarter of a day's wage to a poor family in Khartoum.

Where they do not pay in cash they pay in time spent queuing at a public tap which provides water for a few hours each day, or in diseases caught by drinking or bathing in water from heavily-polluted streams and pools. The problem which is faced most acutely by the poor is that of access to water in adequate quantity, with reasonable convenience, and at an affordable price.

Public water points are frequently damaged, causing wastage of water from leaking taps, and high maintenance costs. This is often a sign that not enough water points have been provided. The provision of more water points may diminish the frequency of breakages due to over-use or vandalism, by reducing the number of households using each tap. A water point serving a

Fig. 9.2 When living on a hillside slum in Lima without any piped water, daily chores are particularly difficult. Photograph by J. Holland, OXFAM.

small group of families (say, less than 40 households) is more likely to be looked after. It may even be possible to levy a water rate from them jointly, whereas larger groups would not be able to organize regular collections.

Neither cost nor consumption is greatly increased if extra street taps are supplied; but consumption increases two or three times when a tap is provided within each house or yard (Cairncross and Feachem 1983).

In many tropical cities the existing treatment plant, pumping mains, and storage capacity are already heavily overloaded so that only an intermittent supply is provided, and there is little benefit in extending the distribution network until the overall capacity of the whole system can be increased.

Meanwhile, a variety of local solutions are possible. Local grants can be offered to householders to install guttering and rainwater catchment tanks, on the grounds of their reduced consumption from the city water supply. Local wells in poor urban neighbourhoods may provide water of good quality if a few basic precepts are observed in their construction. They will normally cost less per person served than an extension to the main city water supply, and certainly be easier to implement. They will also have the virtue of being under the community's control.

An intermediate option is to install a series of small, independent networks of water points, each served by a local source such as a spring, well, or

local sol'n

Fig. 9.3 Boys add potable water they have fetched from a protected source to a drum used for the neighbourhood water supply in a Guayaquil slum, Ecuador. UNICEF photograph by B.P. Wolff.

borehole. These networks can be designed in such a way that they can be connected into the city's overall system, and house connections offered to those who wish them, when the main system's capacity allows.

Rainwater drainage

Without the proper management of stormwater, to prevent flooding and ensure ground stability, the other sanitation measures are practically impossible.

Improvements in rainwater drainage have similar effects to land reform, giving security of tenure and creating conditions for other sanitary improvements. Once secure from the threat of floods, landslides or landlords, the inhabitants will be more ready to invest in improving their own houses, building toilets, and so on. Some years ago, the Sewerage Company of Guayaquil, in Ecuador, calculated the average cost of landfill, to raise the ground level of a household plot above the flood level of the mangrove

Fig. 9.4 Poor drainage is often perceived as the greatest infrastructural problem in poor urban areas such as this *kebele* in Addis Ababa. Photograph by C. Goyder.

swamps on which the city's shantytowns are built. They found that it was no more expensive than the cost per household of the raised pit latrines they had been building. More important, households benefiting from landfill almost always built their own toilets, often better ones than the latrines built by the Sewerage Company. In many cases, they even built new houses.

Rainwater drainage is often the measure given the highest priority by those who live in poor urban neighbourhoods. The importance of this issue was illustrated when in 1982 the Mozambican National Institute of Physical Planning conducted a survey of city councils, asking them what were the most keenly-felt infrastructural problems with which they needed help from the central government. Rainwater drainage was the almost unanimous response, and the evidence is that this reflected strong political pressure from the residents' organizations at local level.

Problems in implementing rainwater drainage include a lack of suitable technical literature and the division of responsibility for drainage between bodies in the city. However, the problems are not completely intractable, and with support from the community some remarkable improvements can be made surprisingly cheaply. An example of successful stormwater drainage in a poor shantytown is the Peixinhos pilot project run by the Prefecture of Olinda in Recife, Brazil (Rego and Cuentro 1984). Precast cement drainage channels, semi-circular in section, and about 20 cm wide and reinforced with coconut fibre, were made at a local level and laid in the ground to create a network running along all the narrow alleys of the neighbourhood and even extending into individual plots. Although the area is nearly flat, care was taken to ensure a gentle gradient of at least 0.3 per cent so that water would run along them, to the nearest discharge point outside the area; in this case, a trunk main of the city's sewerage system.

Water also flows in these channels during the dry season, as they also serve to take away sullage water. The inhabitants are encouraged to pour this directly from their washing basins into the nearest channel on or just in front of their plot. A series of small grids trap debris before it can pile up in large quantities downstream, and each householder is thus rendered responsible for cleaning the debris in his own section of channel. Clearly, careful explanation and organization was required to ensure that these responsibilities would be clearly distributed, understood, and discharged by the residents. The residents had participated in the planning and construction of the system, and the channels extended into their plots, which may have helped to encourage a sense of ownership.

Sometimes an urban neighbourhood may have excellent drainage at local level, but be flooded by water backing up from a major obstruction downstream. This was the experience of the *kampung* improvement programme in Jakarta; drainage along individual streets and paths was greatly improved, but the complete elimination of flooding would have required major engineering works to remove points of constriction on some of the major canals of the city. The cost of such works often amounts to millions of dollars, requiring intervention by the central government.

Refuse disposal

Without adequate solid waste disposal stormwater drainage will not work. Accumulated domestic refuse is the most common cause of blockage of urban drainage channels. Indeed, if refuse is not regularly collected, it is hard to think of where else to put it. It is no coincidence that the successful Peixinhos project mentioned above involved a solid waste disposal component just as important as its drainage efforts.

Unlike other urban infrastructure, the major cost of solid waste disposal is in its day-to-day operation, rather than in the capital investment. Whereas the central government or international agencies can help with finance for capital investments, the operation of a refuse collection system has to be paid for largely for the municipal recurrent budget. Although a smaller amount

of refuse per person is created compared to developed countries, the rate at
which flies breed in the warm climates and the mainly vegetable refuse of
Third World cities means that collections have to be made more often,
usually daily or three times a week (Flintoff 1976). Certainly, the economic
conditions of developing countries demand less capital-intensive systems.
Labour-intensive methods will also reduce the foreign exchange required
(for vehicles and spare parts, for instance), which may be difficult for a local
authority to obtain.

Unfortunately, municipalities often still prefer motorized vehicles where
handcarts or tricycles would suffice. Expensive mechanized composting and
incineration plants are imported and are often not used due to unsuitability
or high running cost (Holmes 1984).

An effective refuse collection service needs good organization, and
depends on the co-operation of the public. If skips or other communal rubbish
containers are not regularly emptied, people will cease to use them. On the
other hand, poor urban communities have been known to make remarkable
efforts, even sweeping their own streets, if they know that a vehicle will
arrive at the agreed time to collect the sweepings.

Poor communities produce less solid waste than the relatively wealthy, but
their refuse is also more dense so that compaction equipment is not required,

Fig. 9.5 An effective solid waste disposal system depends on co-operation and
co-ordination between the community and the municipality. In this *kebele* in Addis
Ababa, small bins are emptied into five skips twice weekly; the skips are emptied once
a month by the municipality. Photograph by C. Goyder.

and it has a higher content of vegetable matter and a lower carbon/nitrogen ratio so that simple composting methods can be used to recycle it for agricultural use. This would be ideal if kitchen gardens were developed (see Chapter 6 for a discussion of kitchen gardens).

Many of the urban poor already earn their living by recycling solid wastes, and often suffer appalling rates of exploitation. They could benefit directly from the income generated by a neighbourhood manual composting unit, or from a guaranteed price for their recycled products. The Peixinhos project showed that composting and even paper-making were viable activities at the scale of a neighbourhood of only 2000 people. In an interesting World Bank project in Medellin, Colombia, assistance was given to scavengers to enable them to buy equipment such as sewing machines or tin shears for making simple products from the waste. Another example is the informal refuse collection programme in Bandung, Indonesia (Poerbo *et al.* 1984).

Sullage disposal

Sullage, or waste water from bathing and washing clothes, and dishes, can often be dealt with by the excreta disposal system, or in many areas of sandy soil or low population density sullage can simply be thrown on the ground without causing harm. However, when the population density or water consumption increases, the limited infiltration capacity of the soil may mean that on-site excreta disposal systems (see below) are not able to cope with the sullage water flows. In such cases some form of drainage network is required; a drainage system for sullage alone, coupled with on-site excreta disposal, is much cheaper and easier to build than a full sewage network.

A system of open channels, as in the Peixinhos example already mentioned, would have the advantage of serving for stormwater drainage as well. It has been suggested that open drains can represent a health hazard (Oluwande *et al.* 1978). It is true that sullage contains faecal matter, and any open drains will, in any case, receive some inflow of sewage from overflowing septic tanks and the like. If stagnant pools are formed *Culex quinquefasciatus* mosquitoes may breed in them and transmit Bancroftian filariasis. But if the drainage channels are too small for children to paddle in, and are designed with a curved invert and kept clear of debris to prevent pools from forming, the risk may be slight. Epidemiological evidence on the subject is scanty, and it would be worth further study, in view of the economy of laying open rather than closed channels.

Excreta disposal

Conventional water-borne sewerage has been found to be prohibitively expensive for any but the wealthiest residents of a Third World city, is almost impossible to install in narrow winding streets, requires a constant supply of large amounts of water for its functioning, and creates frequent problems of environmental pollution. Pollution arises when sewerage systems discharge into bodies of surface water which are used by the population for domestic use or for fishing, and when failure or intermittence of the water supply, or the use of traditional anal cleansing materials, causes blockage.

The World Bank and others have sought and rediscovered several alternatives (Kalibermatten *et al.* 1980). The main on-site disposal options are pit latrines, pour-flush toilets and, where water is available, septic tank systems. In some cases simplified or small-bore sewerage systems and cartage are also worth considering.

On-site technologies   In Addis Ababa, 'ventilated improved pit' (VIP) latrines have been built on an experimental basis by the Department of Environmental Health in a new housing colony constructed on the outskirts of the city entirely by self-help. These are individual family latrines which have been constructed simply, at a cost of around 300 Birr per latrine. UNICEF have funded this pilot project and are hoping it will be developed elsewhere in the city as a part of the PHC programme. The main problem in using this model in certain *kebeles* is that people have been taught to throw excess water down the latrines to avoid the problem of standing stagnant waste water. There is also a tendency for all kinds of rubbish to be thrown down latrines. The decomposition principle will not work if additional waste and water are thrown into the tank. The VIP latrine could be developed for individual family use, especially in congested areas where there is no access for a suction truck and where the family can be taught not to throw any additional waste down the latrines (Goyder 1985).

Composting toilets allow naturally occurring bacteria to 'compost' excreta, rendering it harmless and suitable for use as a fertilizer. It is only likely to be successful in situations where the manurial value of the final product is appreciated. This method has found widest use in Vietnam where the double-vault system is developed. In this system one chamber is used while the other which was previously filled and sealed is composting: by the time the second chamber is full the contents of the first chamber is ready for removal and to be used as fertilizer. The system does require considerable user skill: the rate of filling, liquid content, and addition of other organic matter is critical: urine and sullage have to be disposed of separately. This system has been used in a *kebele* in Addis Ababa, with one double-vault latrine for 15 families at a cost of 3908 Birr (Goyder 1985). Generally, these are not suitable for high-density urban areas: however, they can be upgraded to pour-flush systems (with a U-pipe providing a water seal), using one vault for sullage and the other for excreta: this in turn can be converted to a small-bore sewerage system at a later date.

A septic tank is a covered settling tank which separates and digests most of the solid matter in sewerage. The effluent flowing out of the tank and the sludge accumulating in the tank must periodically be removed. This is an on-site system suitable for households with a full piped water supply, but it will then require a large drainfield which will not be feasible in high-density population areas. In a sense, the key element of the system is the drainfield, not the septic tank.

Off-site technologies   On-site disposal systems are sometimes overloaded, by sullage water or by the large discharges of cistern-flush toilets, or due to the low infiltration capacity of rock or clay soils. On-site disposal can also contaminate the groundwater and so affect adjacent wells (Lewis *et al.* 1980).

Alternative off-site technologies include small-bore sewage which involves the removal of solids in a septic tank and the laying of, say, 10 cm-diameter pipes to remove liquids.

Another off-site alternative is vault and cartage systems. These are extensively used in the Far East. Basically, they consist of a sealed-vault latrine which is emptied regularly by a vacuum tanker: the tanker carries the sewerage to treatment sites outside the town or city. This system has the advantage of flexibility and low initial cost: the size of the vaults will vary with the number of users and the frequency of collection, but usually the latrines can be sited in the house. Where there are several stories it is possible to serve these with a communal vault on the ground floor. This system cannot carry more than 10 litres of water per person per day, so separate sullage disposal systems are needed.

In areas of undisciplined building, the need for access is often not considered and can be a major constraint to the system, which is totally dependent on a well-organized fleet of tankers. It is usually easy to connect the vaults to a sewerage system if this is installed at a later date.

Bucket latrine systems are widely used and, although usually extremely unhygienic, involve the same principle as the vault and cartage system. At its most simple, the excreta is collected from a bucket placed under a normal squatting plate. When the latrine is inside the house a trap-door arrangement gives access to the bucket from the street. The bucket is emptied with a spoon-like dipper into a night-soil bucket and then carried in a wheelbarrow to a collection point. This is usually reckoned as a poor form of sanitation: the manual work involved is highly unpleasant, and spillages and unlawful dumping often occur.

However, until other systems can be offered organized cartage will continue and it is certainly better than no system at all. Relatively minor changes, such as washing facilities for buckets, tight-fitting bucket lids, and personal protection of collectors, can greatly improve performance and hygiene. In the long run the system lends itself to conversion to the vault and cartage system.

Communal toilet   With few exceptions, communal or public toilets are not well maintained. The exceptions are those cases where an attendant is paid to keep them clean and in good order. In Patna, India, he is paid from the revenue of a small usage charge, and distributes a little soap to each user. Lighting and a water supply must also be provided.

However, in most cases the latrine belongs to the householder (or to the landlord) and its maintenance is his responsibility. If an on-site system is used, no public infrastructure is involved. The finance and construction of such excreta disposal improvements is therefore different from the other kinds of infrastructural upgrading, for which there is a single client (a municipal authority, water company, etc.) who can engage a contractor to carry out the construction work. For excreta disposal, the householder is the client; and financing latrines means lending to the householders.

This poses considerable problems to the donor agencies, and although the World Bank and some of its borrowers speak of 'cost recovery' in this sector, most of the schemes proposed so far are unlikely to work. For example, the idea of a mortgage on a latrine, proposed by consultants for Dar es Salaam, is preposterous. When borrowers default, as they certainly will, is the lending agency to repossess their latrines?

Financing   The real problem in financing latrine programmes is a political one. Most latrine programmes for the urban poor which depend on a cash input, in the form of a loan or a grant from a central agency, will never gain sufficient political and financial backing to spread beyond a limited area and serve any sizeable percentage of the urban poor. Governments do not give the sector enough priority, and in so far as the poor can exert political pressure, they have more urgent requests to make than to demand money for latrines. If a programme is to be replicable on a large scale, it will have to be largely self-financing.

This does not mean that governments and municipal agencies do not have a role to play in improving excreta disposal. In fact, cash may not be the principal constraint limiting the people's ability to make their own improvements. Rainwater drainage, land tenure, and shortages of key building materials (such as steel or cement) have already been mentioned. Technical support may also be needed, or the prefabrication of specific components, such as slabs or squatting pans, which require special materials or equipment for their manufacture.

Organizing for sanitation

One constraint which applies to all the five areas of sanitation discussed here is the spatial organization of the community. There may not be enough room for pit latrines, or wide- and straight-enough streets for effective drainage.

An initiative from a central agency needs to enable a voluntary spatial reorganization of the neighbourhood. In some of the shantytowns of Maputo, Mozambique, the central government worked with community organizations to plan a rearrangement of individual plots by which straight streets and access roads were created. Such a project will have little lasting effect without the support of the community. This case illustrated how the residents' preferences may not be obvious to outsiders; several government ministers who visited the area said they thought the plots were too small, but the local people preferred them that way so as to allow wide streets with easy access by fire engines and ambulances.

In the same neighbourhood, a latrine programme is in progress which compares very favourably with the low-cost sanitation programme in Dar es Salaam. Initial planning for both programmes began at about the same time, in the late 1970s. In Maputo, it was found that what people most wanted was a safe but cheap means of covering the pit. The programme therefore centred on a precast latrine slab which could be made in a simple neighbourhood workshop at a cost accessible to even the poorer families (Brandberg 1983). Since the slabs can be sold for profit, their production is a means of income generation, and is in the hands of small co-operatives, many of whose members are women. The co-operatives receive technical support and buy the necessary raw materials from the city council. In this way, several hundred latrines can be built each month, with a minimal injection of capital. Thousands have already been built. Those families which cannot easily raise the cost of a latrine on their own can do so by joining a rotating credit association, a form of organization well established in many poor urban communities.

In Dar es Salaam, on the other hand, a more expensive type of latrine was planned, which most residents would take several years to pay for. Long-term finance was therefore needed, and the project is to be funded by international aid. Planning has taken years, and the programme is still at the stage of building demonstration latrines. Project staff acknowledge that they will be glad if they manage to build a few hundred latrines in the next year or so.

The moral of this story is perhaps that organization and ideas are frequently the most important role for funding agencies, but they must be flexible enough to respond to the felt needs of the community, which may often differ from the expectations of outsiders. Planners often talk of the desirability of community participation in their projects, but the most successful projects tend to be those in which the *planners* first participated in experiencing the problems of the community. From that point onwards projects usually acquire a momentum of their own.

Case study 1   Safai Vidyalaya, Patan: A latrine conversion programme

Based on a paper by Patel (1985).

Until this programme became operational, urban excreta in Gujarat was removed manually by 'scavenger' sweeper-class workers using a bucket and cart system. For the workers this was degrading as well as dangerous from a health point of view. There was also considerable spillage which, combined with difficulties of disposal, meant that raw sewage was often deposited in open drains or rubbish dumps close to human dwellings.

The programme aimed to replace this 'dry' system with 'pour-flush' toilets which are relatively cheap, need little maintenance and infrequent desludging.

The main agencies operating the programme were the municipalities and the voluntary agency Safai Vidyalaya. The municipality was primarily concerned with cleaning up the streets while Safai Vidyalaya Sanitation Institute was originally motivated by the wish to improve the living and working conditions of sweepers and scavengers.

Developing the programme

Because of the high cost of sewerage systems the most common choice was the cheap pour-flush water-seal latrine. Only 1.5 litres of water are needed to flush this type.

The methodology can be illustrated by the case study of Patan city, Gujarat, India, which has a population of approximately 80 000. Ninety-eight per cent of houses are connected to a water supply. At the start of the programme there were around 4000 private 'dug' latrines and 311 public ones. Each latrine serves about 11 persons.

The programme developed along the following lines:

- The government resolution on latrine conversion was discussed at the municipality board meeting.
- A special meeting was held for sweepers to inform them of the changes and to explain that other work (e.g. road building) would be made available.
- A briefing session with technical staff.
- Various meetings with communities at ward level to explain conversion, new by-laws and financing.
- Raising of public awareness through local newspapers, film shows, and training camps.
- A systematic household survey of latrines was carried out and many problems solved on the spot by Safai Vidyala workers.
- Creation of a 'motivating and implementing cell' at ward level.
- Social workers, supervising staff, sanitary inspectors, and engineers were given a special training by Safai Vidyalaya.
- Eight teams of masons and small contractors carried out the work with

assistance from the community. Construction materials were provided by the municipality.

- The cost per household varied between 300 and 650 Rupees: 50 per cent of this was borne by the community.

Achievements

Over the 13 years 1969–82, 4249 latrines were converted and a further 690 new ones constructed. In addition, where housing density made private latrines impractical, communal latrines serving 400 households were built in 16 blocks of the city.

It is worth noting that in Gujarat as a whole over 170 000 latrines have been converted and some 6000 linked to a biogas system. It is clear that the success of this programme relied on a combination of a feasible, cheap technical solution and a sensitive approach to the community involved. Discussion, education, and an ability to sort out individual problems at a household level were essential elements. Lessons can be learnt from the co-operation of a municipality and an NGO which combined organizational ability with resources and sensitivity.

Case study 2   The low-cost sanitation programme of the Orangi Pilot Project, Karachi, Pakistan

Based on a paper by Hasan (1985).

The problem

Two million people live in slums (*'katchi abadis'*) in Karachi. Orangi is one of these slums: it has a mixture of religious groups, poor health conditions, and an infant mortality rate of 110 per 1000. The area is very flat, coastal land. The Orangi Pilot Project is an example of a programme which initially focused upon sanitation but which now encompasses other community development activities.

The target area of the Orangi Pilot Project (OPP) consists of about 4000 acres, contains 3181 lanes, and about 43 000 housing units.

Except for a recently installed water supply system through stand pipes, and the on-going low-cost sanitation programme of the OPP, (begun in 1980) there are no urban services in this area.

There were three major problems in providing a sanitation system to the squatter colonies here:

- The local authorities do not have the necessary finances for constructing a conventional sewage sysstem.
- The cost of urban services, as developed by the local authorities, were nine times the actual cost of the labour and materials required for such development. Users in squatter colonies cannot afford to pay these charges.

- Community participation and organization was difficul to achieve because of self-interested 'community leaders'. The most homogenous unit in the area was the group of lane residents.

Programme aims

Keeping the above factors in mind, the low-cost sanitation programme of the OPP aimed from the very beginning at discovering:

- alternative sources of finance from within the community available before the development work itself was undertaken;
- alternative low-cost methods of implementating the development;
- alternative methods of community participation.

GOALS

To achieve the above objectives it was necessary to study the sociology, technology, and economics of the community's solutions to the sanitation problem and see if the OPP could build on them.

Pre-OPP solutions to the sanitation problem

Before the OPP's low-cost sanitation programme was accepted by the residents of Orangi, three solutions to the problem were commonly used by the people:

- the bucket latrine—emptied into open drains after flooding the streets;
- soak-pits—which were expensive, filled quickly, and drained into the streets;
- individual sewerage lines—very expensive, uncoordinated, and substandard.

The OPP felt that if an effective lane organization could be developed, and if the right kind of technical support and tools could be provided and the lane residents trained to use them, then an underground drainage system could be developed in Orangi.

OPP's low-cost sanitation programme

Three concepts are central to the understanding of the sanitation programme of the Orangi Pilot Project. First, community participation. The sanitation programme is a result of a need to develop lane organizations in Orangi and not the other way round.

Secondly, there is a need to move away from previous technology. Standard engineering technology and implementation procedures, the product of the traditional client–engineer–contractor relationship, have to be constantly modified to suit the new system where the user, organizer, and implementor are one, and often they have little or no technical knowledge or artisan skill.

Thirdly, the programme has implications for social dynamics. In the process of organization and participation in development work, changes are bound to occur in the community. These changes will result in a redefining of relations with the local government, the scope of future development work undertaken, and its manner of implementation.

OPP's methodology for development of lane organization and technical support

At an early stage the OPP social motivators held meetings of lane residents and explained, with the help of slides, posters, and pamphlets, the benefits of the OPP low-cost sanitation programme. In addition, the motivators stressed that without the formation of a lane organization the OPP could not give any technical assistance. Lane managers were elected, who, on behalf of the lane, formally applied to the OPP for assistance. There is no standard structure for the organization that emerges and it varies from lane to lane.

The OPP technical staff then survey the land, establish bench marks, prepare plans and estimates (of both labour and materials), and hand over this data to the lane managers.

The lane managers collect the money from the people, call meetings to sort out any social problems which may occur due the undertaking of this work,

Fig. 9.6 The Orangi Pilot Project in Karachi encouraged residents of the lanes to organize themselves so that sewerage lines could be laid. Photograph by A. Hasan.

receive tools from the OPP, and make arrangements for carrying out the work. The OPP staff supervise this work. At no stage does the OPP handle or manage the finances of the people. This is an important factor in the success of the programme.

To organize people on a large scale without the simultaneous existence of smaller organizations is impossible. The organizational unit therefore, was limited to a lane. As such, physical planning was done for one lane at a time. Now after 4 years of work, these smaller organizations are well established and the unit of operation has increased, with planning now being done for entire neighbourhoods.

Education and evaluation

Initially, substandard work was done in the lanes by the people. This was due to a lack of central supervision and a failure to appreciate the value of OPP technical advice and education. In mid-1982 there was a lull in the programme. As a result, an evaluation of the concept, design, and implementation procedure of the project became necessary. This showed several main weaknesses.

First, according to the initial designs of the OPP, the sewerage and excreta were discharged into the open out-flow area '*nullah*' which became a serious health hazard. Secondly, because of lack of water to carry sewage, many lines were clogged up and had to be cleaned out.

To overcome these two problems, it was decided to place a one-chamber septic tank, or '*haudi*' as it is known in Orangi, between every connection and the sewage line. This prevents the solids from flowing out into the drain. The size and design of the *haudi* was determined not according to any engineering standard but by its cost to the user.

The manhole covers posed another problem. They were originally expensive and needed artisan skills. The designs were modified to make small holes and heavier covers which prevented people lifting them to put rubbish into the sewers. The cost was cut to one-third of the original.

To improve the quality of the concrete mix, small metal sheets for mixing concrete have been proposed so that the concrete should not be mixed on the ground. In addition, proper measuring boxes for aggregate are now given along with other tools. Focusing on lane meetings, a vigorous education programme involving talks, posters, and pamphlets resulted in the residents acting as watchdogs to ensure that concrete was adequately cured by masons and contractors.

These various improvements were implemented slowly over a period of time so as not to undermine the confidence of the community.

The OPP has engaged a full-time plumber who supervises the work in the lanes. It has also commenced a programme for the training of masons in the theory, design, and implementation of the low-cost sanitation programme.

Fig. 9.7 A one-chamber septic tank, or '*haudi*', is used outside every house in the Orangi Pilot Project, Karachi. Photograph by A. Hassan.

The names and addresses of the trained masons are given to the lane residents so that they may employ one of them for their work. Monitoring of the results of the work done by these masons shows a marked improvement in quality.

Criticisms of inability to scale up

Sanitation engineers were very critical of the work in the lanes being carried out without a master plan, which involved a complete survey of the area. They feared that it would not be possible to proceed beyond the lanes that bordered the open natural '*nullahs*' and that the programme would end there.

However, OPP felt that if the people, now organized at lane level, were educated regarding this problem, and if the local councillors could be made to get involved in seeking a solution, they could proceed further. The technical solution posed no serious problem. So they decided to work towards it.

Due to lobbying by the Orangi Pilot Project and organized lane pressure, the Karachi Municipal Corporation have appointed consultants to prepare a master plan for the sewerage system. The engineers are certain that work done by people through Orangi Pilot Project's advice can be integrated into this.

The project develops

In order to promote the concept of secondary drains in the area of each elected councillor, a physical survey was undertaken by students who were helped by the OPP Orangi residents. This plotted the slope of the land, number of houses, number of lanes, and existing '*nullahs*'. The survey could have been carried out by professionals in a short period of time. But it was carried in this way for four reasons:

- To promote an understanding of sewerage system among the people, without which no further community work was possible.
- To take the concept of development through local participation to the professional colleges and universities.
- To involve the councillors by making their area a unit of research, and to arm them with facts and figures about their area and with a vision of a better future.
- To educate OPP's workers and the people of Orangi who, in turn, would pressurize the councillors.

By involving the people and their representatives in the survey, groups of lanes are now coming forward to have their secondary drains built. On the basis of the information OPP have provided, the councillors are asking OPP to prepare plans and estimates so that they may pressurize the municipality into financing the people's schemes.

Some statistics and costs

The project is clearly a success. Out of a total of 3050 lanes in that part of Orangi, over 1200 have already built up their drainage system. Of these, 458 have operated with Orangi Pilot Project's advice. In addition, about 60 secondary drains have been, or are being, laid. The cost of the OPP sanitation programme at three levels (the sanitary latrine in the home with a *haudi*, the underground sewage pipe line in the lane, and the underground concrete pipe secondary drain) comes to less than Rs. 1000 per house. The people with OPP advice have done work worth about US$1 100 000. If the local government had carried out this work, it would have cost them about US$4 400 000 and the quality would have been inferior to the work of the Orangi Pilot Project. For every Rs. 100, or US$6.5, spent by the people, the Orangi Pilot Project has spent about Rs. 10, or US$0.65. This includes capital expenditure as well as administrative costs. The figures will keep reducing as the work increases.

People's awareness and the change in people–local government relationships

The awareness by the people has brought about various changes in the relationships of the Orangi Pilot Project, the people, the councillors, and the municipality, with the sanitation programme and with each other.

The population now understands sanitation technology, appreciates good quality work, knows costs, and therefore will not permit kickbacks and profiteering. This led to people wishing to get their own secondary drains built rather than pay the high cost of municipality work. The sanitation programme has now become a movement.

More and more, the Orangi Pilot Project organizers are becoming involved in the supervision of work rather than organization, which the people do for themselves.

The demand for tools is also increasing: to reduce the work load on staff OPP have prepared charts which tabulate costs and quantities of materials and labour for every item of work. Armed with these facts, the lane people can make estimates themselves, if they wish to.

Councillors get grants from the municipality for certain specific development works in their areas according to municipality specifications. Now, however, people have forced the councillors in two areas to spend these funds on laying underground sewage lines using their own cheaper methods. The main reason given by the local bodies in Pakistan for not being able to provide sanitation to low-income areas is cost. The Orangi experience, is an example of a low-cost, community-based system of urban waste disposal. However, the problem can only be tackled in this manner if authorities accept this decentralized approach. This process seems to undermine administrative and executive authority and develop the power of the people. Under certain conditions, therefore, it may fail initially to acquire the respectability necessary for its integration into the official planning process.

This general approach to physical planning is now being applied to other problems in Orangi, such as cheap new housing and improvements to old housing.

The new social organizations and motivation created by the low-cost sanitation programme is now being used by the Orangi Pilot Project to further its welfare programmes for women. Thus, the lane communities developed in the OPP programme make ideal entry points for exploring other improvements in the health field.

Comments and conclusions

The general introduction and case studies above give no global blue-print for urban sanitation. The solution chosen in Karachi rejected small-bore sewerage which, at least superficially, appeared to be the most appropriate system. Various factors such as geography, population density, water supply, cost, and cultural background mean that individual solutions have to be sought for each situation. However, some general points can be made:

- Involvement of the community at the early stages of planning of any scheme is likely both to enhance its chance of success and greatly reduce the cost of any system.

- For maximum cost effectiveness, government, municipality, and NGOs should work together so that resources and understanding of the community are maximized.
- Unlike private facilities, communal latrines will only operate hygienically if workers are paid to supervise and keep them clean.
- Systems of excreta disposal must include an understanding of present and projected water supply to the area. As a corollary, water supply must be planned in conjunction with sullage removal.
- Technological options vary with population density, and the number of people using the same latrine: under high-density situations the only alternatives are vault latrines, cartage, and sewerage systems.
- Where population density is increasing, the technology of choice should be adaptable to these high-density options.

10

Housing

It seems clear that we cannot achieve Health For All by the year
2000 without Shelter for All by the year 2000.

Munro, 1986

Housing policies and the urban poor

Many efforts have been made over the last several decades to examine the
specific relationship between housing and health. Hinkle and Loring (1977)
and Mason and Stephens (1981) provide excellent reviews of this work.
Housing, although an important ingredient, is just one of the multiplicity of
socio-economic factors that influence health, and it has proved almost
impossible to attribute a specific health condition to housing alone. Any
modification of the physical environment will have the maximum effect on
health only if accompanied by other socio-economic and environmental
improvements. Nonetheless, there is sufficiently strong evidence relating
certain aspects of the residential environment as a cause of ill health and this
has affected housing design and planning decisions (Stephens *et al.* 1985). A
summary of housing design features and the diseases that they may help to
overcome are given in Table 10.1.

Large-scale housing developments can have a detrimental effect upon
health. As described in Chapter 4, the higher costs of new housing can reduce
a family's disposable income, resulting in unintended effects such as a
poorer diet and less adequate health care. However, improved housing with
security of tenure is often a priority of the urban poor.

Governments and development agencies have responded in a variety of
ways to the enormous demands for housing and urban infrastructure arising
from rapid urbanization. During the 1950s and 1960s illegal shanty towns
around official cities were considered by most governments as temporary
problems which could be solved by public housing programmes and by
clearance and relocation projects. In many countries in the 1960s large-scale
housing programmes, based on the construction of conventional dwellings,
were started for the first time. With few exceptions, these programmes had
little impact. Often the number of new dwellings constructed as part of
public housing programmes did not even keep pace with the number of
housing units destroyed in attempts to eradicate slums and shantytowns
(Hardoy and Satterthwaite 1985*a*).

Although mass evictions and bulldozings do still occur, governments
increasingly recognize that these simply destroy the only kinds of accommo-

134

Table 10.1 Housing design features and the diseases that they may help to overcome

Design features	Diseases combated
Strong association	
Adequate supply of water	Trachoma, skin infections, gastroenteric diseases
Sanitary excreta disposal	Gastroenteric infections, including intestinal parasites
Safe water supply	Typhoid, cholera
Bathing and washing facilities	Schistosomiasis, trachoma, gastroenteric and skin diseases
Food production means	Malnutrition
Control of air pollution	Acute and chronic respiratory disease, respiratory malignancies
Fairly strong association	
Ventilation of houses (in which there is smoke from indoor fires)	Acute and chronic respiratory diseases
Control of house dust	Asthma
Siting of housing away from vector breeding areas (including stagnant water)	Malaria, schistosomiasis, filariasis, trypanosomiasis
Control of open fires, protection of kerosene or bottled gas	Burns
Finished floors	Hookworms
Screening	Malaria
Some association	
Control of use of thatch material	Chagas's disease
Rehabilitated housing	Psychological disorders
Control of heat inside the shelter	Heat stress
Adequate food storage	Malnutrition
Refuse collection	Chagas's disease, Leishmaniasis

Source: Stephens *et al.* (1985).

dation that most of the urban population can afford. Thus, upgrading projects for slums and squatter settlements and 'sites and services' projects are becoming more popular (Satterthwaite 1985).

Table 10.2 gives the common characteristics and problems of the most common forms of owner occupation available to the urban poor.

In upgrading, the occupiers of houses or shacks in illegal or unauthorized

Fig. 10.1 Public housing programmes such as this in Lagos, Nigeria, were thought to be the answer to illegal shantytowns in the 1960s. They proved to be very expensive and are now occupied by middle-income groups. Photograph by Salami, WHO.

Fig. 10.2 Advertising luxury in Lagos. The 'housing policies' of many governments are to keep illegal settlements out of sight—either moving them to remote outskirts or even erecting billboards so they cannot be seen from main roads. Earthscan photograph by M. Edwards.

Table 10.2 The most common forms of 'owner occupation' open to low-income groups in Third World cities

Types of 'owner occupation'	Common characteristics	Problems
Building, house, or shack in squatter settlement	As city grows and number of people unable to afford a legal house or house site grows, illegal occupation of land sites on which occupants organize construction of their own house or shack usually becomes common. Advantage of what is usually a cheap (or free) site on which to build—although as the settlement develops, a monetized market for sites often develops and land sites can be expensive in better quality, better located settlements. The extent to which households actually build most or all their house varies a lot; many lack the time to contribute much and hire workers or small firms to under-take much or all the construction.	Lack of secure tenure; settlement often subject to constant threat of destruction by government. Lack of legal tenure inhibits or prevents use of site as collateral in getting loan to help in construction. No public provision of water, sanitation, roads, storm drainage, electricity, schools, health care services, public transport—or even where government does so, this is well after settlement has been built and is usually inadequate. Poor quality sites often chosen (e.g. subject to flooding or landslides) since these have lowest commercial value and thus give the best chance of avoiding forceful eviction.
Building, house, or shack in illegal subdivision	Together with housing built in squatter settlements, this represents the main new source of housing in most Third World large cities. Site is bought or rented off landowner or 'middleman' who acts as developer for landowner. Or where customary tenure is still common, access to a site through the permission of the appropriate chief who acts for the 'community'. Govern-	Comparable problems to those above except land tenure is more secure and landowner or developer sometimes provides some basic services and infrastructure. The site is also usually planned (although so too are some squatter settlements). The better located and better quality illegal subdivisions are also likely to be expensive. If city's physical growth largely

Table 10.2 *continued*

Types of 'owner occupation	Common characteristics	Problems
	ments often prepared to tolerate these while strongly suppressing squatter occupation. Often relatively well-off households also organize their house construction on such illegal developments. As in squatter settlements, the extent to which people build their own houses varies a lot.	defined by where squatter settlements or illegal subdivisions spring up, produces a haphazard and chaotic pattern and density of development to which it will be very expensive to provide infrastructure and services.
Building, house, or shack in government site and service/core housing scheme	An increasing number of governments have moved from a concentration on public housing schemes (which were rarely on a scale to make any impact) to serviced site or core housing schemes. Very rarely are these on a scale to have much impact on reducing the housing problems faced by lower income groups.	Public agency responsible for scheme often finds it impossible to acquire cheap, well-located sites. Sites far from low income groups' sources of employment chosen, since they are cheaper and easier to acquire. Extra cost in time and bus fares for primary and secondary income earners can make household worse off than in squatter settlement. Regulations on repayment, building schedule and use of house for work or renting room often bar many and bring considerable hardship to those who do take part.
Invading empty houses or apartments	Known to be common in a few cities; overall importance in Third World not known.	Obviously insecure tenure since occupation is illegal. May be impossible to get electricity and water even if dwelling is connected.

Source: Hardoy and Satterthwaite (1985*b*).

settlements can be given land tenure plus infrastructure and basic services, and their efforts to improve their shelters are supported with credit and cheap building materials.

The problems faced by those who rent accommodation (whether in an inner-city slum or in an illegal development) are more complex. The most common forms of rental accommodation open to the urban poor are analysed in Table 10.3. Upgrading programmes often give too little attention to the fact that many of the poorest individuals and households are tenants or sub-tenants or simply occupants of cheap boarding houses on a day-to-day basis. Programmes have to address their needs in terms of improved living environments (and perhaps strengthen their rights) but without increasing accommodation costs.

RENTERS & asst

Hardoy and Satterthwaite (1985*a*) point out that if squatter invasions or illegal subdivisions are in fact providing most of the sites for new housing (which is the case in an enormous number of cities in developing countries), it is tempting to suggest that a government's main role should simply be to give legal tenure to the inhabitants of such settlements after they have formed. Subsequently, infrastructure and services can be extended to them without the government getting involved in substantial land acquisition and development programmes. This is more or less what has happened in Bogota over many years. While public authorities have acted to suppress squatter invasions, they have taken a more tolerant attitude over illegal divisions of land. A high proportion of Bogota's population lives on illegal subdivisions. Over time, city authorities have given legal tenure to well-established 'pirate' subdivisions and even extended basic services to them (Carroll 1980). But the long-term cost of only reacting to and eventually legitimizing what is already happening can be very high. As Carroll notes, 'the social costs of the uncontrolled development of *pirate* subdivisions are probably large. These include installing services in unsuitable places such as steep slopes or flood zones, building service networks for inefficient lot layouts; providing transportation services to inaccessible areas; revising street and utility construction programmes to take account of unauthorized development; and coping with erosion, pollution and the destruction of natural amenities as a result of development in ecologically sensitive areas.' (Carroll 1980).

Housing in integrated community development

Truly integrated community development programmes will address the housing needs of the population. The Undugu Society, whose community health programme was described in Chapter 8, operates several programmes, one of which is the Kitui Village housing project aimed at the improvement of slum housing (Ongari and Schroeder 1985).

Kitui is a slum settlement in Pumwani Division in Nairobi. It consists of approximately 1000 families, mainly from the Kamba tribe. They live in

Table 10.3 The most common forms of rental accomodation open to low-income groups in Third World cities

Types of rental accomodation	Common characteristics	Problems
Renting room in subdivided inner-city buildings (tenements)	Usually the most common form of low-income housing in early stages of a city's growth. Buildings usually legally built as residences for middle- or upper-income groups but subdivided and turned into tenements when these moved to suburbs. Advantage of being centrally located so usually close to job or income-earning opportunities. Sometimes, rent levels controlled by legislation. Certain Third World cities never had sufficient quantity of middle/upper-income housing suited to conversion to tenements to make this type of accomodation common.	Usually very overcrowded and in poor state of repair. Whole families often in one room, sometimes with no window. Facilities for water supply, cooking, food storage, laundry, and excreta/garbage disposal very poor and have rarely been increased or improved to cope with much higher density of occupation brought by subdivision. If subject to rent control, landlord often demanding extra payment 'unofficially'. Certain inner city areas with tenements may be subject to strong commercial pressures to redevelop them (or their site) for more profitable uses.
Rented room in custom built tenement	Government-built or government-approved buildings specially built as tenements for low-income groups; sometimes publicly owned. Common in many Latin American cities and some Asian cities: and usually built some decades ago. Some quite recently constructed public housing estates fall into this category.	Similar problems to above in that original building never made adequate provision for water supply, cooking, ventilation, food storage, laundry, excreta and garbage disposal. Inadequate maintenance common.

Rented room or bed in boarding or rooming house	Often most in evidence near railway station or bus-station though may also be common in other areas, including illegal settlements. Quite common for newly arrived migrant family or single person working in city to use these. Single person often hiring 'bed' within a dormitory or even, in some, hiring a bed for a set number of hours each day so more than one person shares the cost of each bed. Usually relatively cheap and centrally located.	Similar problems to above in terms of over-crowding, poor maintenance, lack of facilities. A rapidly changing population in most such establishments prevents united action on part of users to get improvements.
Renting room or bed in illegal settlement	In many cities, rented rooms in illegal settlements (see Table 10.2 for description) represent a larger stock of rental accomodation than in tenements which were legally built (see above). May take form of room or bed within room rented in house or shack with *de facto* owner-occupier; may be rented from small- or large-scale landlord even though it is within an illegal settlement.	Problems in terms of quality of building and lack of infrastructure (paved roads, sidewalks, storm drainage . . .) plus site often ill suited to housing as in squatter settlements and in illegal subdivisions (see Table 10.2). Also, insecurity of tenure which is even greater than for *de facto* house/shack owners.
Renting plot on which shack is built	The renting of plots in illegal subdivisions or renting space to build a shack in some other person's lot or courtyard or garden are known to be common in certain cities; its extent in these and other Third World cities is not known.	Similar problems to those listed above in terms of insecure tenure and lack of basic services and infrastructure. Additional burden on household to build, despite no tenure and no incentive to improve shack.

Source: Hardoy and Satterthwaite (1985b).

semi-permanent, one-roomed, mud-walled houses roofed with corrugated iron sheets or paper cartons, plastic, and scrap metal. Most of the houses have only one door and one window.

There used to be about 350 families in igloo-shaped shelters made from plastic. But after a fire broke out, which destroyed about one-third of all the plastic igloos in June 1983, the Undugu Society helped villagers with materials to build their present shelters. Unfortunately, other people from outside made use of this confusion and built their houses in Kitui Village, tripling the number of families living there and making the place incredibly overcrowded and destroying the unity and the sense of community in the slum.

The case study of the urban community development in Hyderabad, presented in Chapter 8, also provides a good example of including housing in a broader development programme.

The 'felt needs' of poor urban communities often include improved housing as a priority. For example in Kebele 29, Addis Ababa, nearly every sector of the community recently expressed a need for improved housing (see Table 10.4). Thus, housing forms a particularly good starting point.

Programmes sometimes specifically use housing improvement as an initial activity for community organization and then incorporate health activities afterwards. This is what has happened with the NGO Human Settlements of Zambia.

Fig. 10.3 The housing of the urban poor takes many forms. Here a family lives under a tent made of blankets in a Bombay street. Photograph by Virkud.

Fig. 10.4 Large drainage pipes form a convenient shelter for street dwellers in Bombay. Photograph by V. Virkud.

Case study   Human Settlements of Zambia, Lusaka

Based on a paper by Jere (1985).

Background

By 1971, 30 per cent of Zambia's population lived in urban areas. In 1976 the population of Lusaka was 950 000, and squatter settlers were building four units for every one built by the authorities.

The Second National Development Plan recognized that: 'although squatter settlements were unplanned, they nevertheless represent assets both in social and financial terms. The settlements needed planning and services, and the wholesale demolition of good and bad houses was not a practical solution' (Jere 1985). This important decision meant that local authorities had to embark on programmes of upgrading selected squatter settlements by supplying piped water, gravelled roads, garbage collection, sanitary sewer systems, and street lighting. To achieve the desired programmes, the local authorities encouraged local community organizations. One voluntary organization involved in community work is the Human Settlements of Zambia (HUZA) which works in Lusaka.

HUZA

HUZA was formed in 1982 with the object of promoting self-help and self-reliance for social development in the field of housing. It is funded by

Table 10.4 Proposals in order of priority as selected by all six
committees* of Kebele 29, Addis Ababa, 1986

Priority	Score
(1) Improvement of housing	120
(2) Building latrines	120
(3) Improved supply of drinking water	83
(4) Improvement of feeder roads	72
(5) Flood protection	71
(6) Relief assistance for the poor and destitute	67
(7) Kebele office building	56
(8) Kindergarten	50
(9) Improved drainage	44
(10) Kebele meeting hall improvements	42
(11) New co-operative shop	42
(12) Income generating projects	34
(13) Clinic	25
(14) Recreation centre	24
(15) New grinding mill (for corn)	19
(16) Study room for poor students	8
(17) Street lighting	7
(18) Community shower	3

*Six Committees—executive, health, development, social, women, and youth.
Source: unpublished data.

church and aid organizations and by local contributers. Jere (1985) points
out that the role of HUZA has been facilitated by the nature of community
participation in the Zambian context. The political system is defined as a
'one party participatory democracy' and the upgrading project in Lusaka
was administratively structured along similar lines to the party. Jere suggests
that 'the local authority project management structure is unique for its
flexibility and willingness to accept and integrate suggestions from community leaders and residents' (Jere 1985, p. 8). Interestingly, a similar point was
made about the urban community development programme in Hyderabad,
presented in Chapter 8.

The local authority project
The spread and growth of squatter housing based on traditional and rural
models lacked the necessary urban amenities considered necessary for safe
living for the residents of these areas. They used water from shallow wells,
often badly contaminated, resulting in numerous cases of intestinal diseases.
There were no schools, health services, proper markets, or roads to enable
service vehicles such as ambulances to reach the interior of the settlements.

The residents had reached the limit of doing things for themselves. However, there were many positive things about these areas. It was for this reason that the government introduced the policy of squatter upgrading. The policy's objectives were to address these shortcomings. The policy recommended the following services in squatter upgrading projects:

- The project in Lusaka provided facilities and services to 29 000 households with a total population of 180 000.
- Piped water based on one standpipe for a group of 25 households. The system was designed to meet predictable demand for the next 30 years.
- Tarred and surfaced roads with access to each group of houses. However, additional feeder roads were to be provided through self-help by the people themselves.
- Urban health centres with a limited number of admission services.
- Schools to provide for the educational needs of the children.
- Improvement of markets or the construction of new ones where none existed.
- Improvement of sewage systems and storm-water drainage.
- Community centres and pre-school facilities for children.

At an early stage, the local authority attempted to ensure that residents were made aware that the project was meant to support their own efforts and to avoid creating the feeling that the project was meant to solve all their problems. Rather, as the National Housing Authority team put it, it should be seen as a 'division' of responsibility between the 'community' and 'authority' in the upgrading process: to promote and encourage self-help projects and other self-reliance programmes chosen by the community itself, as was the case before the areas were designated for upgrading. The community organization needed in such a programme is enormous and the local authority therefore needed the help of voluntary organizations such as HUZA.

The sharing of objectives—housing and employment

Poor housing conditions and related services are the result of poverty. It is for this reason that HUZA and the local authorities decided to integrate the projects to meet both shelter and economic needs. The following are jointly promoted as one project:

- self-help housing and squatter upgrading;
- skills training projects and income generation;
- health, nutrition, and kitchen gardening activities.

multi-prong approach

Self-help housing   Through self-help, households have been able to build their houses cheaply: otherwise home ownership would have remained a dream. The economic promotion programmes have generated building materials which are priced much more cheaply; including bricks, doors, and roofing materials.

Skills training   With improved skills, families are able to produce goods for their own use and any surplus is sold to the immediate community. However, the supplementation of family income is the main objective, both in cash and in kind. With increased income they also are able to invest small savings into house improvements. Some of the residents, after gaining skills, have formed production units. Youth groups produce building materials used in self-help construction work. Women's groups have been established to make soap, soya-coffee, candles, and school uniforms which bring in a good income and make a small contribution to the national economic development.

Health and nutrition education and kitchen gardens   Mothers are taught how to grow some of their own food and how to preserve food. The important role women play in the development of their families is stressed. Health education is an important element but it is realized that such developments are threatened by the lack of the means to practice it (for example, in promoting the use of oral rehydration, lack of income may preclude the purchase of sugar and salt for the ingredients). Therefore, skills are taught which can be used in productive informal employment (e.g. tailoring, knitting, making soap, candles, cooking oil).

Helping the poor to help themselves

Jere (1985) emphasizes that the call on the urban and rural poor to undertake self-help projects should be made with care. Doing so, without helping the people to help themselves can easily be interpreted by the poor as an academic exercise. The poor are only able and willing to be mobilized, organized, and involved in development programmes and projects when they perceive that benefits will accrue to alleviate their suffering. The poor have no time or energy for academic exercises, but are only concerned with their everyday survival. They have shown enterprise in using their very limited resources. In helping them, care must be taken so that:

- False expectations are not created by community workers which can lead people to believe that all the problems in their areas will come to an end because of the presence of the workers or of their agencies. In this respect it is important to be very precise as to the level of resources that may be brought into the community.
- Only what is absolutely necessary should be made available, in order to assist in activating dormant or under-utilized community resources.
- Nothing should be done for the community that the community can do by itself.
- Use of local buildings, tools, and human resources must be encouraged when ever possible.

The HUZA programme also emphasizes that community-based appropriate technologies and productive skills demonstrations should be undertaken in conjunction with the residents. These can be very important educationally. Many everyday items, such as soap, candles, and cooking oil, are considered difficult to produce yet, if introduced as part of community education and demonstrations, the amount of self-satisfaction among community residents and the level of self-confidence generated is surprising. They become stepping stones towards other attempts at income generation. These learning exercises have proved very important. Currently, HUZA works with over a dozen community groups on income-generating activities. These successes have played a major role in popularizing such income-generating activities in Lusaka. They have proved that, with consistent organization and community education, much more can be achieved.

Communication

In Lusaka's urban community development projects much experience has been gained in the use of appropriate modes of communication in community mobilization. Carefully-prepared facts, information, details, and objectives of the project formed the basis for communication, with emphasis being placed on the need for increased local participation, self-help and self-reliance by the communities. Wherever possible, local musicians and artists were used at community meetings. They sang promotional songs which were recorded and then played on national radio. Large attendances at community meetings consequently became a common feature of the project. A local drama group worked with HUZA to communicate health education. 'The group is based in a low-income community and represents local dreams, aspirations, traditions and hopes.' (Jere 1985.)

Comments and conclusions

The 'self-help' approach used by groups like HUZA can be used by governments as their excuse for doing little or nothing. For many poor households 'self-help' in terms of having to construct part (or all) of their house is an enormous burden. Their survival often depends on more than one of their members (including children) working very long hours. The fact that many income-earning activities are in the so-called 'informal' or 'unrecorded' sector means that these people are often characterized as 'unemployed' or 'underemployed'. In reality they are 'over-employed' in terms of the hours worked, but they receive grossly inadequate incomes from this work. Thus, government initiatives must understand the range of housing needs among 'the poor', and differences in the range of skills, resources, and incomes that different households can bring to bear on helping to solve their own housing problems. Different individuals or households with similar incomes will have different needs in terms of location, size and form of dwelling, and extent to which payments can be made for a house and services (Hardoy and Satterthwaite 1984).

Programmes to upgrade low-income neighbourhoods rather than bulldoze them has considerably increased knowledge and experience in recent years. The Kampung Improvement Programme in Indonesia, initiated in Jakarta and later extended to many cities (Karamoy 1984), upgrading projects in Dar es Salaam, Tanzania, and in Lusaka, Zambia, have demonstrated that the environment in low-income areas can be considerably improved at relatively modest per capita costs. Generalizations about what should be done are more difficult, in that the specific forms poor housing environments take are very varied. Improving water supply and provision for sanitation and garbage removal is an almost universal need: so, too, is provision of health care and education.

Perhaps the single most useful action authorities can take is the granting (or regularizing) of land tenure in illegal settlements, for this removes the perpetual fear of forceful eviction and gives households the security they need to improve their houses. Once an illegal settlement is legalized, there is a stronger case for the public provision of basic infrastructure and services.

Fig. 10.5   Landscaping in the city. *Source*: D. Brunner.

11
Integrated approaches

> The slum dweller is not a parasite . . . Slum dwellers are pro-
> ductive citizens with the right to participate in decisions that
> affect them. They have been treated as simply a problem, an
> unsightly settlement. But they are a vital part of the city, and it
> is very important that the city provide them with all the neces-
> sary services.
>
> Bill Cousins, UNICEF Senior Urban Advisor

The need for integration

The previous chapters have focused on different interventions which have
approached health problems through different health-related activities.
Sometimes municipalities or voluntary agencies emphasize health education
and preventive and promotive health programmes without considering the
importance of other health-related activities which directly affect the health
status of the target group. Health education, immunization, nutrition
education, etc. might be emphasized, but little effort is made to relate these
to the provision of safe drinking water, maintenance of a clean and hygienic
environment, adequate shelter, income generation, etc. In this chapter we
focus upon initiatives which have adopted a much more integrated approach
from the beginning, that is, where the priority is for the total development of
the community, of which health improvement is an essential part.

Until recently this integrated, 'community development' approach seems
to have had little impact upon urban policymakers. Urban community development has potential for the building of systematic linkages between physical improvements, social services, and people's participation. Cousins and
Goyder (1986) for example, have emphasized the need for a community
development approach in the integration of physical improvements. The
reconstruction of a slum community requires the willing co-operation of the
residents so that they give up part of their plots or part of their buildings so
that straight lines may be made or drainage lines laid. Often drainage or
housing is installed without consulting local residents. The attitude of the
people then is 'you built it, so you can maintain it.' This is one explanation
for the maintenance problems that affect many housing and slum improvement projects. 'When community workers indicate when and where the
people are ready for these improvements and act as liaisons between the
people and the engineers, the changes are more likely to be welcomed, understood and long lasting.' (Cousins and Goyder 1986.)

149

Fully integrated programmes often have more success in generating community self-help, together with significant improvements in self-awareness and community function. Eisenberg (1980) describes entering an area within Comayaguela, Honduras, which had an integrated programme:

'Having walked along dirt roads marred by foul-smelling cesspools and past wretched hovels that saddened my heart for their unhappy inhabitants, I was unprepared for the sudden transition to clean passageways between brightly painted houses that bore the names of their inhabitants clearly marked on their doors. The homes, mostly of wood and corrugated zinc, were no larger than those I had just passed, but painted and clean; they exuded pride and cheer despite their poverty and crowding.'

This change in the community was partly the result of an integrated programme which focused upon the expressed needs of the people:

'The initial stage was one of data gathering by direct observation and by interviews, in order to understand the characteristics of the physical environment and the needs expressed by the members of the community. The second stage was to be a programme of education and health promotion to raise the level of consciousness about community problems. The third was to assist in organizing goal-directed groups within the community targeted at self-help activities to deal with the expressed problems. The final step was to be an evaluation of what had been accomplished, an evaluation undertaken jointly by community members and health specialists.'

Activities of the programme included the organization of weekly meetings of the community (there were 300 people in the area); the development of a women's group which covered issues such as the use, distribution, and preparation of food; health care; personal hygiene; household management; and 'mental health' advice (such as how to respond to epilepsy, alcoholism, drug addiction, and delinquent behaviour). A water reservoir was constructed and a small industry of paper bag manufacturing was created. An adult literacy programme was developed with the help of a local school teacher.

The following two case studies take a similar integrated approach with the first one indicating what can happen when an integrated programme narrows its focus.

Case study 1   Cheeta Camp in Bombay—BUILD

Based on a paper by Pinto (1985).

Cheeta Camp is a resettlement slum in Bombay city consisting of a population of 60 000 comprised mainly of migrants from outside the state of Maharashtra. Initially, this population was allotted a piece of marshy land, undeveloped and devoid of all basic services. Gradually, lanes were built and

Fig. 11.1 The people of Bombay's Cheeta Camp are still trying to make a new life for themselves after eviction at gun-point from their well-established slum community at nearby Janata 10 years ago. Photograph by A. Charnock.

small-scale commercial activities emerged—grocer's shop, a timber depot, vegetable and fruit vendors, restaurants, etc. Private health clinics were set up and the area was also provided with a police station.

The majority of the people in Cheeta Camp are Muslims and a large number of them (71 per cent) were agriculturists prior to their migration to the city. Nearly half of them have been in Bombay city for more than 20 years. Fifty per cent of the people are illiterate, while another 25 per cent are educated up to primary level. Although there are education facilities up to the seventh standard in the vicinity of the slum, the school dropout rate is very high. Their health status is poor with a high prevalence of leprosy and tuberculosis. Gastrointestinal ailments, such as diarrhoea, are prevalent. The public latrines and water supplies are inadequate.

BUILD is a voluntary organization which has been working with these people since 1976. In 1977, BUILD established its health centre and four sub-centres and organized a group of community health workers. At first the agency had provided the inhabitants with housing. The community workers helped the community to demand and get services like a water supply and sanitation facilities. Hence, from the start the agency was involved in a wide range of health-related activities. The concern of BUILD was total development of the community of which health improvement was only a part.

Mode of action

There were about 15–20 community workers selected from the community, who attended to minor illnesses and referred the serious ones to one of the four sub-centres run by a nurse. If the illness was complicated, the patient was referred to the main health centre run by doctors. There was a social worker at the central level who attended to activities related to prevention, and promotion of health aspects such as sanitation and drinking water. If problems relating to, say, sanitation were encountered by the community health workers, these were referred directly to the social worker who then discussed the issue with the community, organized them, and helped them to demand action from the municipal corporation. The health staff and social worker therefore functioned as a team.

Services improve—social development declines

Over a period of time, people gained most of the basic services, such as housing, roads, water supply, and public latrines. At this stage, the agency felt that the broader health-related activities were coming to an end. This led to the isolation of the health care programme from other activities, resulting in the disintegration of the previously comprehensive social development programme.

Most of the present activities of the agency are now restricted to a narrow definition of health care. The vital loss to the programme was the breakdown of the referral system between the community health workers and the social worker. Thus, the agency's integrated community development programme ended up becoming a curative health service.

In the BUILD health programme the patient's first contact is with the health worker and then with the nurse at the sub-centre. Only if required, contact is established with the doctor at the main health centre. Patients often prefer to approach private practitioners in order to have direct contact with a doctor. This also happens when the agency's opening times are inconvenient for the people. The credibility of the agency has diminished because there is now less need for its involvement in securing a wide array of services for the community. The agency failed to identify a new role for itself in the context of the changing needs of the community. The additional factor which reduced its credibility, even as a health service agency, was the lack of effective support and referral services (Pinto 1985).

[handwritten margin note: probs & pgm]

Case study 2   Kebele 41, Addis Ababa

Based on papers by Goyder (1985) and Teferra (1985).

Conditions in Kebele 41

Addis Ababa is administered by elected committees in a system of urban dwellers' associations called *kebeles*. There are 284 *kebeles* with an average population of 5000 which make up the 25 higher *kebeles* or *kefetegnas*. There is a city council, or municipality, with a mayor, which supervises the affairs of the city in the *kefetegnas* and *kebeles*. The urban health sector is administered by the Addis Ababa Regional Health Authority under the Ministry of Health (MOH).

Two integrated urban upgrading projects exist in Addis, each with a considerable emphasis on community health. The more extensive of these projects in Kebele 41, Kefetegna 3, was supported by the NGO, *Redd Barna* (Norwegian Save the Children Fund), with additional support from other agencies, including OXFAM UK. The project began in 1981 and it was intended that *Redd Barna* should phase out its assistance in 1986, with the *kebele* taking over responsibility for most of the programme. *Redd Barna* has begun supporting another project in Kebele 13, Kefetegna 21. The emphasis of this project is more on upgrading environmental conditions and assisting with income generation, while the strength of another project run by the NGO, Concern, in Kebele 37, Kefetegna 4, has been the maternal and child health programme.

With a population of 4101, Kebele 41 is located in one of the poorest and most congested parts of Addis Ababa, identified by a World Bank survey as the poorest of seven *kebeles* where the Bank was considering supporting upgrading activities. The 788 houses standing in 1981 were crammed into 6.7 hectares. There is no self-sufficiency in Kebele 41, as its residents, a high number of whom are unemployed, must survive totally within a cash economy and in a social setting where many of the traditional supports of the extended family have broken down. During the past four years the project has mobilized local labour and resources and by channelling activities through the existing *kebele* structure, and has developed an integrated development programme based around three components; health, physical upgrading, and community development (see Figs 11.3 and 11.4). The project has emphasized that preventive health cannot be isolated from such activities as water

Fig. 11.2 The typical housing in Kebele 41, Addis Ababa. Photograph by C. Goyder.

supply, sanitation, income generation, education, and training (Teferra 1985). Activities within the three major components are as diverse as preparing legumes to sell, suction-cleaning of latrines, the provision of a health post, repairing houses, and organizing literacy meetings. Below we examine just some of these activities.

Improving the environment

Physical upgrading is one of the three main components of the programme. Environmental conditions in Kebele 41 at the start of the project were appalling. A great deal has been achieved to raise these standards during the past four years and visits to surrounding *kebeles*, where there has been little attempt to improve the environment, provides a striking contrast to the relatively good conditions now predominating in the *kebele*. By the end of 1984 13 communal latrines had been constructed at a cost of 3000 Birr per six-seater latrine (2 Birr = US$1.0). Some 54 latrines have been reactivated and repaired, and some of the most dangerous latrines closed. The cost of upgrading existing latrines is about 2000 Birr, depending on the condition of the original pit, and these are then turned into communal latrines wherever possible. In addition, 24 individual latrines have been suction-cleaned and

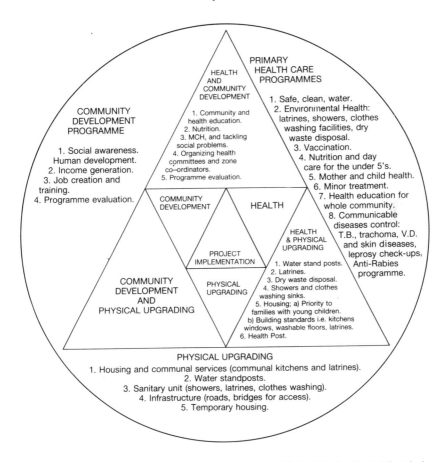

Fig. 11.3 The holistic approach to community health in Kebele 41, Addis Ababa. *Source*: Teferra (1985).

repaired at a cost of about 600 Birr per latrine, with families providing the labour requirements. Communal latrines are assigned to specific families. A maintenance rota written on each latrine door makes it easy for the sanitary guards or the CHWs to supervise latrine cleaning. It has been found that the programme works well and it is now managed by the *kebele*, although the project is still paying the salaries of the sanitary guards and the costs of latrine suction.

Three water points have been constructed by the project and one of the existing two standposts has been repaired and three taps added. Water is sold at a cost of three jars for five cents and is available for residents from three surrounding *kebeles*. A washstand, shower and latrine complex have been

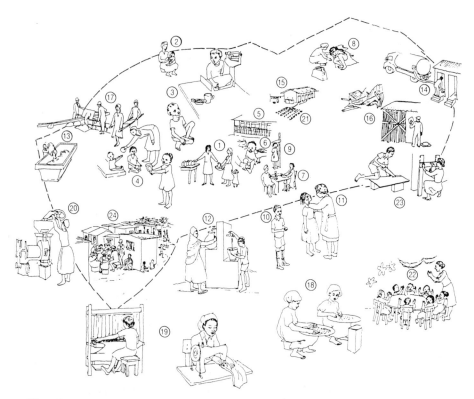

Fig. 11.4 The activities of the community health programme in Kebele 41, Addis Ababa. *Source*: C. Goyder.

PRIMARY HEALTH PROGRAMME AND ACTIVITIES

NUTRITION PROGRAMME

A Catholic Relief Service
 * Dry ration supplementary food distribution
   (dried milk, sorghum, oil) to mothers
   and their children 0–5 yrs old ①
  *Food preparation demonstration for mothers
   of under 5s. Weaning food. ②

B Nutrition Rehabilitation Centre
 * Feeding of marasmic and underweight children ③
 * Demonstration of nourishing food preparation
 * Hygienic food handling
 * Concentrated health education on prevention
   of diarrhoea and vomiting and
   the use of oral rehydration salts

C Day-care room for marasmics ④
   43 square meter room accommodating
   22 malnourished children while their
   mothers are employed in the Income-
   generating scheme.

HEALTH POST ⑤

A Vaccination of children under 5 yrs & pregnant
   women ⑥

B Ante-natal check-up ⑦
C Emergency delivery by traditional birth
   attendant ⑧
D Post natal check-up
E Well baby check-up ⑨
F Family planning
G First Aid (24 hours)
H Films-health and development education
I Referral
 * High risk antenatals
 * Disabled children for expert advice
 * Corrective operations for polio victims ⑩
 * Deaf and dumb placement in institutions
J Communicable diseases control
 * T.B. Mass check-up for under 14s
 * Trachoma mass check-up ⑪
 * Voluntary V.D. check-up for women
 * Skin diseases check-up

WATER

3 communal water distribution points
each with 10 taps. ⑫

SANITATION

A Environmental hygiene
Waste disposal system: 42 waste bins
emptied into 5 skips (13)

B Latrines
* Excavation, renovation and reactivation
of old latrines
* Building new communal latrines
* Vacuum suction to empty latrines (14)

C Personal hygiene
Shower and washstand unit: (15)
* comprises 8 showers
* 4 clothes washing basins
* 2 latrines, and washing lines

PHYSICAL UPGRADING
HOUSING AND INFRASTRUCTURE

A Building new houses to replace condemned ones (16)
B Major and minor repair of houses
C Relief housing (Temporary shelter during building works)
D Bridge building and associated road improvements (17)
E Soil-cement block making

COMMUNITY DEVELOPMENT
INCOME GENERATING

A Food processing
* 100 sq meter corrugated iron shed
* 30–50 mothers employed preparing
legumes for sale (18)

B Garment making
132 sq meter concrete block shed
20–30 unmarried mothers

* Weaving cotton for blankets, curtains, shawls
* Sewing (19)
* Printing and dyeing childrens clothes

C Grain mill (20)
A reasonably priced service for Kebele 41
and people of neighbouring kebeles

D Vegetable garden
* Cultivated by mothers in food processing
and children 7–14 years
* Irrigation from shower unit (21)

EDUCATION AND TRAINING

A Kebele school kindergarten (22)
B Literacy programme for children unable
to attend school
C Sponsorship of school drop-outs (23)
Training in carpentry, plumbing, electrical
and other skills
D Juvenile delinquency prevention
Trainig 7-14 yr olds to make easily marketable
products and to work in the vegetable garden
Disabled children are included in this

COMMUNITY AWARENESS AND DEVELOPMENT (24)

A Zone co-ordinators (community leaders)
meeting once a week.
B Health committee meetings every other week.
C Development, literacy and education
meetings every week.
D Sports and culture, community shop and
social welfare committee meeting every week.
E General community programmes progress
report and evaluation meetings held
quarterly.

Fig. 11.5 Community latrines are built and looked after by the residents of Kebele 41, Addis Ababa. Photograph by C. Goyder.

constructed at a cost of 25 000 Birr. People pay 10 cents for the use of the shower and 25 cents for washing clothes. Clothes-lines are also provided. A similar shower constructed in Kebele 13 makes use of solar power to provide hot water.

It is intended to use the waste water from the unit in Kebele 41 to irrigate a nearby vegetable plot which is being cultivated by unemployed youth and by some of the poorest women, who are processing food grains in an income-generating project. Further details about the sanitation schemes used in other projects in Addis can be found in Chapter 9.

The dry waste disposal programme is now working well. Forty small bins are emptied by the community into five skips which are placed at convenient access sites for the municipal refuse trucks. Small bins are emptied twice weekly by the community and the skips are emptied about once a month by the municipality. This programme had some initial communication problems. The first bins were distributed before the community really appreciated the need for the programme. The bins, which were large and painted a fresh white, were considered far too useful for rubbish and the community preferred to use them for storing grain! This incident provided a reminder to the project staff to ensure that no programmes were introduced in future without health education, discussion, and participatory management by the local community. The waste disposal programme has cost 8600 Birr, and the programme is supervised by the sanitary guards who also sweep lanes and keep the *kebele* clear. There are monthly sanitation campaigns, and there have been regular film shows and discussions on all aspects of environmental health and personal hygiene. One problem has been the shortage of good films for this purpose, especially films with direct relevance to local conditions. Diarrhoeal diseases are a major health problem. However, the situation has improved considerably since the first two months of the project, when 15 children died during an outbreak of diarrhoea and vomiting.

Other environmental improvements within the programme include building new houses to replace condemned ones, minor self-help repairs to housing, and the improvment of access roads (see Figs 11.3 and 11.4).

Linking nutrition and income generation

As the figures demonstrate, the primary health care activities in Kebele 41 are numerous. It is, perhaps, the problem of malnutrition which best underlined the necessity for an integrated programme.

In Kebele 41 there is still considerable malnutrition. There were 98 children registered with the Nutrition Rehabilitation Centre (NRC) programme in January 1985. These were children whose weight had not increased above 70 per cent of weight for age over a period of three months. Many children weighed less than 70 per cent.

Eighteen children who had failed to gain weight over a two-year period were placed in a day care unit. It was the mothers themselves who pointed out to the project staff that their children were constantly sick and failed to gain weight because they had no option but to leave children unattended while they worked as daily labourers. Most of these mothers had no other support. Now the mothers are earning around 2 Birr a day in the food-processing plant which the programme set up and their children are fed and cared for six days a week. In addition, mothers participate in food demonstrations which emphasize the cooking of food currently on sale in the *kebele* shop because food shortages were preventing the sale of traditional foods such as teff and wheat. There is regular nutrition education and cooking demonstrations for mothers of children attending the NRC.

Overall, great progress has been achieved in the Kebele 41 project but the prevailing conditions of extreme poverty, food shortages and high food prices, cultural inhibitions to change, and overcrowded living conditions will continue to prevent people from achieving the type of health status found among some self-sufficient rural communities.

The Kebele 41 project shows that a great deal can be done to improve environmental standards even in crowded urban areas. Furthermore, a considerable effort has been made to raise income, especially among the poorest members of the community. Priority is always given to helping families with young children, and an attempt has been made to raise income within the *kebele* by providing income-generating projects such as the *kebele* grain mill. This should help to provide income to finance projects when *Redd Barna* have withdrawn their support, but it seems likely that the *kebele* will always require some support since the community remains very poor. The *Redd Barna* project has received criticism because it has been expensive (see Fig 11.6 and Table 11.1). It is fair to say that the initial capital costs of the project have been high but that the recurring costs of the health component are not expensive. The future role of the project must now be to make good use of the resources that have been developed within the project to serve a larger community (see Ch. 15 on scaling up and scaling down).

Comments and conclusions

The complex mix of factors affecting health in a poor urban environment make integrated programmes particularly attractive. Often the felt needs of a community will focus upon improving the physical environment rather than health services. With an established dependence upon existing curative health services such as hospitals, private practitioners, and pharmacists, a health programme which focuses solely on preventive health is unlikely to succeed. The two case studies from Bombay and Addis Ababa were both involved in a wide range of health-related activities. However, the programme in Bombay witnessed improvements in basic services provision and

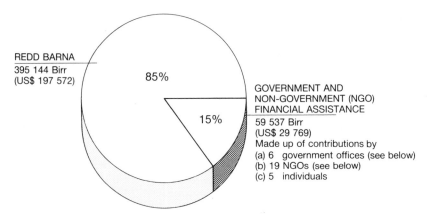

Fig. 11.6 Financial support from government and non-government organizations in Kebele 41 health programme component, July 1981–December 1984. *Source*: Teferra (1985).

Note:

(a) **Government offices:**
   Addis Ababa regional health—including EPI, Lidetta clinic, and Teklahaimanot clinic
   Dodola Health centre
   Ethiopian Nutrition Institute
   Ministry of Health—including St. Pauls hospital
   Municipality of Addis Ababa—including Rabies Control and Sanitation Dept.
   TB centre

(b) **NGOs:**
   ALERT
   American Embassy
   British Council
   British Save the Children Fund
   Catholic Clinic
   Catholic Relief Services
   Christian Relief and Development Association
   Ethio-Italian trachoma control
   Ethio-Swedish children's hospital-Polio Clinic
   Ethiopian Red Cross
   Family Planning Association
   The Ford Foundation
   German Culture Centre
   Indian Embassy
   Norwegian Scouts
   OXFAM
   Swiss Embassy
   UNICEF
   Vianelle Printing

Table 11.1 Financial allocation between components in Redd Barna's integrated urban development project, Kebele 41, Addis Ababa (totals given in Ethiopian Birr: 2 Birr = US$1.0)

				Health component		
		Amount spent on health component				
Year	Total for 3 components*	Total (%)		Nutrition (%)	Water & sanitation (%)	Health post (%)
1981	325000	38856	(12)	24	76	—
1982	374100	137284	(37)	23	31	46
1983	467410	110800	(24)	16	59	25
1984	477079	108204	(23)	11	61	27
Average	411077	98786	(24)	19	57	33

*The three components are health, community development and physical upgrading.
Source: Teferra (1985).

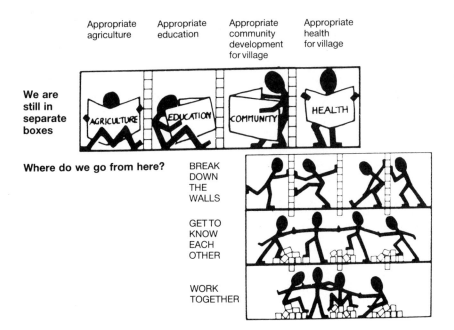

Fig. 11.7 The need for an integrated approach. *Source*: TALC.

then decided to become more specialized. The programme became a curative health service and did not compete with existing curative services. It therefore lost credibility.

Community participation is facilitated by the existence of an integrated programme. As a number of countries have shown, notably China, the potential of the population for improving its own health and living conditions is enormous. When the community is fully involved, collaboration between sectors for integrated action is often greatly simplified.

12

Health services

Variety of health services

Although there is usually a heavy concentration of health services in urban areas, particularly hospitals, the urban poor are not well provided for (see also Ch. 5). These health services face a heavy demand but access by the poor is very difficult. This encourages private, traditional, and informal health care to flourish as alternatives to government services. There is often a profusion of curative health services offered by private and non-government organizations, particularly at hospitals and outpatient clinics, whereas it is usually left to national and local governments to be concerned about health promotion and preventive services.

However, where this is not the case, it is commonly the non-governmental organizations and religious missions that extend their activities into these areas. The provision of health services has been a common entry point for social programmes organized by non-government and charity organizations, such as religious missions, and only more recently has there been a growing concern that hospitals also become involved (for instance see Aga Khan Foundation 1981; Macagba 1984; Hardie 1985; WHO 1987).

Urban people are usually more politicized than their rural counterparts, which makes health programmes a useful point of entry. Neither the ruling powers nor the community feel threatened by health programmes, which, at least initially, appear to be politically neutral. However, when health programmes begin to extend their activities and begin to tackle the fundamental causes of ill health, and particularly if they encourage the development of community participation, it is inevitable that a greater political awareness will develop. Activities such as lobbying for better environmental health conditions and facilities can, in circumstances where injustices exist, lead to mass community actions of a more frankly political nature, which then become a powerful threat to local authorities and national governments. For such reasons there are many circumstances when health services are organized from non-health institutions like churches and missions. Such services give daily contacts with people and are potentially, therefore, a powerful means of health promotion and communication. The problem is that these exchanges may be limited to hospitals, clinics, and mobile health units, with the community playing a passive role and with the services organized from a purely technical viewpoint. There is thus a failure to analyse the causes of ill health and to consider broader measures to improve health, such as water supplies, housing, education, and income generation (see Chs. 9, 10, and 7, respectively).

Thus, health activities in urban areas must involve an interaction between central services and the community itself. The interaction of these two elements is crucial: failure of many health activities is commonly due to inadequate involvement of the community from the early stages, or conversely a failure of the services to support community initiatives. The increasing use of community health workers, with responsibilities to both the services and the people has been one response to this situation (see Ch. 13).

Health programmes may fail to expand to other areas of community development because they are commonly inundated with demands on their curative skills, and because they have a limited desire or mandate to do so. Ministries of Health may not take kindly to their staff becoming involved in political and economic issues, and private services are concerned with their

Fig. 12.1 A health promoter from the Colombia health delivery programme enquires about health problems of a family in the Bogota suburb of El Codito. Photograph by P. Harrison, WHO.

patients and making a profit. In certain circumstances NGO programmes may have the advantage in that their functions are not so clearly defined and they can branch out into other activities.

The following two case studies illustrate these points well. The first outlines the development of community-based health services amongst poor urban areas in Chile at a time of severe political repression. The second illustrates the need to expand health services so that they take a much broader view of the causes of ill health and are organized jointly *with* community organizations and not *for* the community.

Case study 1   Fundacion Missio, Santiago, Chile

Based on a Paper by Mayer (1985).

The public health service in Chile

From 1950 to 1973 Chile developed a national system of social medicine. In 1952 the national health service and the national system of social insurance were legally sanctioned which made the government responsible for the health care of the majority of the population. Both workers and employers contributed to and were members of the health insurance system. Those who were unemployed or who had no stable employment had free medical treatment provided by public hospitals. The organization of the public health service was the responsibility of the government and great importance was attached to preventive medicine.

In 1979, after the overthrow of the socialist government, the military government implemented a complete change and health care had to be organized through the free market. The government withdrew more and more support from the public health care system. Thus, the poorer sectors of the population received less and less health assistance. Other changes have been:

- The national expenditure on health care was reduced drastically, officially from 3.6 per cent to 2.3 per cent in 1980.
- About 8000 health worker employees were dismissed.
- In 1979 the hospitals were forced by the law to become up to 80 per cent self-financing.
- People in the national health insurance system now have to pay part of their health care, in addition to their contributions.
- Dental treatment has been almost completely restricted to private practice.
- All drugs and supplies have to be purchased on the market, making them more expensive and resulting in shortages.

In addition it has been estimated that more than one-half of the work force in the low-income areas of Santiago are unemployed or underemployed.

Only those who are employed are now entitled to the benefits of health insurance.

The Fundacion Missio considers that each society has as a fundamental goal the achievement of good health for its population, particularly when health is conceived as a state of physical, mental and social well being. The missio must, therefore, be involved in all aspects of the quality of life affecting the '*poblador*' (inhabitants of low income settlements) and their human rights. However, in the present situation their programme can only be palliative and one day the government health sector will have to take on more responsibilities.

The programme developed from modest beginnings. In 1974 a small health care unit was established in one of the low-income areas in the northern sector of Santiago. Workers from the unit trained the local population in preventive and curative medicine and attended those patients who could not go to public or private health care institutions. To make sure that only the really needy—those who had no health insurance scheme—attended this health care unit, promoters were trained to select patients and to provide medical advice. In addition, assistant nurses were trained to carry out simple curative work and later to undertake more complicated activities. By late 1984 more than 400 voluntary promoters were trained and most are still active. Their main functions are in medical care and community health education activities.

The health programme now covers the low-income areas of the municipal districts of Conchali and Renca in the northern part of Santiago, where there are about 350 000 people. Fundacion Missio has built or taken on responsibility for seven health units.

Activities

Psychiatric care is now being included because mental illness has been identified as a major problem. Dental treatment was included in the 1985 programme, since over 90 per cent of people have dental problems.

The medical care programme covers people:

- without access to the health insurance system and those whose health insurance does not provide for free attention and cannot pay private fees;
- rejected by government services, particularly children, who need urgent attention.

The health centre issues drugs to the patients treated by the Missio clinic and to those from other centres who cannot afford to purchase them.

Community health education and training

A fundamental part of the Missio's work focuses on the 'promoter's' course, whereby health groups are provided with basic knowledge on health, and some techniques and procedures to give primary health care to the community. Each of the health centres is a centre for the further education of the promoters.

Assistant welfare workers are chosen by their respective neighbourhoods and form a network of people responsible for the health conditions in their neighbourhood. They do their work in conjunction with the Fundacion Missio. Their task is to communicate to the population their rights and how they can obtain their legitimate medical services. In addition, they decide which people in their neighbourhood can be treated in the Missio health units. Their training includes information about the national system of social insurance; the diagnosis, prevention, and treatment of the main diseases; group dynamics; and the means of analysing the local health situation.

Health instructors are voluntary. They teach about diseases and their prevention, and plan and carry out campaigns of preventive medicine together with the personnel of the project. Normally, health instruction is given through lectures and courses held for local communities, neighbourhood groups, parents, or Christian parishes. Among the topics covered are environmental hygiene, child development, nutrition, early diagnosis of diseases, and people's rights in the national health system. In addition, the health instructors organize care for alcoholics, drug addicts, and delinquents.

Assistant nurses, or assistant medical aides, help in providing medical care for the population. They assist the doctors and nurses in their work by talking to patients and giving simple medical treatments. In the community they administer first aid in their neighbourhood. Their training is practical and includes first aid, treatment of injuries, use of drugs, common diseases and their treatment.

Presently, Missio is working with 44 health groups in seven areas, involving approximately 360 volunteers.

The community groups are expanding their activities to include intersectoral activities in Pincoya, El Cortijo, Renca 2, and El Salto. A major focus will be on the real and perceived role of women in low-income areas.

The decline in services is aggravated by a general deterioration in the economic position of the poor, with unemployment rising. As a result alcoholism, prostitution, and malnutrition are all increasing. Consequently in 1980, the Fundacion Missio restated their aims, acknowledging the Christian responsibility to alleviate the degrading chastisement, the inhuman suffering and miserable life of millions of Latin American people experiencing high infant mortality rates, poor housing and ill health. They also resolved to do what was possible to preserve human rights.

Methods of working

The most important concepts are:

- Teamwork—the technical teams meet fortnightly and the health groups weekly. Both are democratic, with their agendas decided by themselves, so that they are able to programme their own activities.
- Group autonomy—the technical team supports the health groups until they can achieve total autonomy but does not seek to run them.
- Health groups—these are responsive to the needs of the community.
- Unpaid volunteers—these are supported with donations and activities like bazaars, raffles, musical evenings, teas, etc.
- Co-ordination with the health sector—the Fundacion Missio co-ordinates closely with other organizations in the community, such as hospitals and clinics run by other agencies.

In summary, the work must be based on mutual self-respect rather than on authoritarian or paternalistic structures.

The future

The programme realizes that the assistants' work is only a palliative in an overall situation of great necessity, and that it is the government's responsibility to provide this service. Missio sees its main functions as the education and promotion of health so that the potential capacity of the people can be developed during a time when this is often suppressed. It is hoped that the community will become aware of its health condition and, based on critical analysis, will organize itself for its own improvement so that some of the problems may be solved.

Case study 2   An integrated community-based programme for squatters in Kuala Lumpur, Malaysia

Based on a paper by Yusof (1985*a*).

Background

Some characteristics of the squatter settlements in Kuala Lumpur were described in Chapter 3.

The squatter settlements have arisen spontaneously and often very rapidly over a period of months. The size of the settlement usually varies between 300 and 1500 houses containing between 1000 and 8000 people. Settlements tend to grow rapidly in numbers and establish a leadership structure, so that the fledgeling settlement can resist being dismantled by the authorities. In order to survive, the settlement must find influential patrons who will protect their interests. These are usually members of a political party, who trade their support for political allegiance. These patrons may use their influence to obtain the basic necessities for survival, such as water stand-pipes, sanitation, garbage clearance, drainage, and roads. As the settlements gain in strength and self-assurance they become more demanding for these and other services, so that a conflict develops between them and the City Hall.

Because there are no clear guidelines, the authorities make compromises, giving the squatters some services but not too many. While the squatters do not have titles to land, there is no clear policy on either slum clearance or upgrading.

Unlike some other urban squatters, the Kuala Lumpur communities have a high level of employment and are considered a vital source of unskilled labour. Forty-six per cent occupy the service categories such as barber, tailor, domestic help, cook, launderer, parking attendant, postman, labourer, etc.

The total population of the squatter community is now estimated at around 1.25 million people. The dependency ratio is high, with 52 per cent of the population being children under the age of 15 years. The settlements tend to divide on ethnic lines, i.e. Chinese (76 settlements), Malaysians (58 settlements), and Indians (14 settlements).

Sang Kancil—an intervention programme

This programme sprang from increasing awareness of the problems of squatter settlements and has now spread to 15 settlements in Kuala Lumpur. In 1978 the City Hall held a consultative seminar, with the co-operation of UNICEF, at the National Institute of Public Administration. The objectives of the seminar were to identify the problems and needs of children and families in low-income areas of Kuala Lumpur, and to look at better ways of tackling the problems. It was realized that children were the ones most at risk and that any programme to help them must also help the care-providers (mothers).

The problems were summarized as:

- Communicable diseases: a combination of crowding, low immunization rates, and poor sanitation gave a high infection rate. Polio was common, and infestations were common.
- Frequent pregnancies: fertility rate was high with close spacing of children and a rapid population growth which the slum could not contain.
- Psychological stress: it was felt that the breakdown of the extended family and the support it gives put greater stress on both parents and children. This is aggravated by the needs of both parents to work, so that children were often neglected, resulting in high truancy rates and drug addiction.

Problems experienced

The initial thrust of the programme was to introduce MCH services to the urban squatters, but a compromise was made between the felt needs of the community leaders and advice from the services. One of the main problems was the identification of the real leaders of the community as against those individuals who were motivated by personal advancement and the wish to please authorities. The leadership structure varied considerably from political activists, those who had become leaders through a strong personality (charismatic leaders), religious leaders, old people, *tenghulus* (appointed by the government), and elder midwives. Listening to the community has a role in validating information received from more formal sources or from informed 'experts'. The development of Sang Kancil has illustrated this clearly. It was only after listening to the community that the emphasis of entry was changed to pre-school education and income-generating activities. This was in line with the community's felt need that health was not a priority. The practice of listening is of equal importance in deciding the time selected for clinics and in the understanding of local customs.

In developing the programme, difficulties were experienced in identifying the most needy people and in preventing the strong taking advantage of the weak and poor. To this was added the problem of identifying leaders in populations which were rapidly changing, making continuity difficult to achieve.

An attempt at intersectoral co-ordination

At government level too, there have been problems with different ministries exhibiting a reluctance to co-operate, since the departments do not like surrendering power. Experience from the Kuala Lumpur intersectoral 'Nadi Programme' has provided some valuable learning experience on the psychology and mechanics of intersectoral co-ordination at the multi-agency level. Far from being successful, it has, nevertheless, helped planners and future programming. One major obstacle identified is the reluctance of government agencies to be 'co-ordinated'. To the head of department, it seems to imply surrendering one's territorial imperative to another agency. Such an attitude is understood easily enough because the need to be visibly successful at the agency level will influence future funding and promotion. Very few heads of department will relinquish control of their portfolio. The most visible form of intersectoral co-ordination begins at meetings and, for most, it will end there.

There is also a need to understand the process of setting up a co-ordinating mechanism. Kuala Lumpur's experience has indicated that a multi-agency approach with 20 or 30 ministries is counter-productive. Co-ordination is best achieved by slow accretion based on existing projects on the ground, where the contributory agency can share a visible stake in the success of the programme. For example, in the Sang Kancil MCH programme, a close and symbiotic relationship was achieved with the National Family Planning

Board. The latter has provided provided a family planning nurse to work together with a City Hall MCH nurse.

The attitude and approach of government officials to communities is also critical. Government officials are often viewed as authority dictators, while the squatter communities, realizing that they are illegal settlers, are defensive in their attitudes. Under such conditions a free flow of ideas and active contributions by both parties is almost impossible.

Political affiliations of squatter settlements further complicate the issues for both ministries and the community. For instance, certain political groups and their supporters will boycott the health clinic if they are excluded in the initial planning stage. Where the political patron belongs to the opposition party the situation is even more difficult to resolve. It is impossible to bypass these leaders as they are then likely to destroy the project.

Leaders at all levels expect some reward for their support of any programme. This may be limited to praise and accolades but often involves something more substantial, such as placing close family members in charge of income-generating programmes, pre-school and health activities. Often this leads to the poor, high-risk group being excluded from the various services which were originally intended for them. While there may be no way to abolish this practice, experience shows that this can be minimized if the implementor and other influential members of the community co-ordinate the poor together, so that they monitor each other's performance.

Strategy for mobilizing community participation

Yusof (1985b) summarizes the lessons learnt from this programme:

- Keep the project simple and start with something that gives quick results. Invite surrounding squatter settlements to witness what has happened (this is the approach taken by an NGO in Addis Ababa, see Ch. 11).
- Be flexible in objectives and aims, and avoid a tight time schedule. Remember, development must go at the pace of the community and not the project. Try always to build up the confidence of the community and always keep promises.
- Avoid direct confrontation. Be prepared to give way initially.
- Choose motivated people who mix well with both those who are poor and the highest authority.
- Keep track of the project by developing very simple evaluation systems which the community itself can understand.
- Publicize the projects through government and international agencies who may have more credibility with government than the squatter community and thus be able to help ideas to gain acceptance.

Evaluation of the projects

In 1982 a questionnaire-based study was carried out in four squatter communities, two with the Sang Kancil centres and two without (Yusof 1985b).

Both communities in the programme area had antenatal care uptake quadrupling over the period. Uptake of family planning services has also increased, although more slowly. It is hard to measure the ultimate impact of the programme because many related factors exist. However, one clear indicator is the lower instance of anaemia in pregnant women in the programme area as compared with the control. Moreover, in the family planning field, spacing between marriage and first birth was rising in the project area towards a two-year interval while in the controlled areas it still remained at one year.

Comments and conclusion

It has been suggested that programmes using health as an entry point, are less likely to diversify than other types of programmes. While this may be true in some cases, the evidence of the case studies given at the workshop and, in particular, the two in this chapter seem to contradict this. In Kuala Lumpur, pre-school playgroups and income generation are now part of the project.

It is, of course, true that health inputs do need and are expected to have a strong service component, i.e. immunization, drug supplies, etc. which all call upon resources not available within the community. This, in turn, brings delays and sometimes conflict where services and communities interact. However, this process may, in the long run, be beneficial as urban community development must come to terms with the authority vested in the relevant ministries and local government.

In addition, health projects are usually not self-sustaining. No poor community can afford all the expenses of their own health service, although clearly they may have to bear a large proportion of the load. Once again, this sharing of the financial burden necessitates an interaction between centre and community. It is possible that the reluctance to diversify a programme is caused not by the nature of health interventions themselves, but by the health professionals who are responsible for the implementation. Where the performance of doctors, nurses, and other health professionals is paternalistic and limited to their technical skills, community development is unlikely to blossom. It is important, therefore, that the right sort of help from professionals is used in urban community health programmes. Those involved need to have a sound understanding of epidemiology and causes for urban ill health, which necessitates a wider understanding than that provided by medical science alone.

The experience of Sang Kancil, the Santa Fe Foundation in Bogota (Rocuts 1985), and others points to the advisability of carrying out small sample surveys before planning health interventions (see Fig. 12.2). If these are well planned and include open-ended questions and interviews, they will establish whether the views expressed by community leaders are in fact

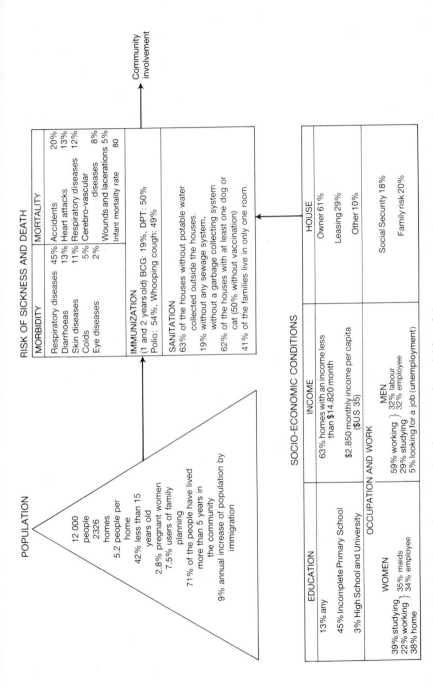

Fig. 12.2 Health situation and associated socio-economic variables of the communities living in the surroundings of the Santa Fe de Bogota Foundation—Sanitary Census. *Source:* Rocuts (1985).

reflected among the population they represent. In other words, a correct community diagnosis can be made (see Ch. 3). It is disturbing to find that both large and small health interventions, while claiming to involve the community, do not take advantage of these simple community survey techniques.

Often a health intervention will start with something simple and acceptable which produces rapid and visible results. While this is quite acceptable and even advisable in the short term, it is important that the community and service personnel that are involved in the programme continually analyse what they are doing in relation to the underlying problems. Surveys therefore, should not be limited to the pre-project period but should be an integral part of monitoring the programme, so that the satisfaction or otherwise of the community with the project can be assessed.

13

Urban community health workers

A range of workers

Community health workers (CHWs) are usually seen as the people most likely to make primary health care (PHC) accessible to nearly everyone; who understand local communities and are responsible to them; who can relate to traditional healers as well as to *modern* medicine; whose costs may be shared between the community and government; and who are also important in political processes and development at the local level. CHWs are seen to be critical in primary health care, but they need to be differentiated from other health workers such as nurses, medical assistants, midwives, and sanitarians, who are clearly a part of the health services.

While the rationale for community health workers is generally accepted, initial enthusiasm is being tempered by caution. Many questions have been

CHW

Fig. 13.1 A community health worker calls on a slum family in Guayaquil. UNICEF photograph by B.P. Wolff.

175

raised about their function, selection, training, and utilization (Vaughan 1980; Heggenhougen *et al.* 1987). CHWs must operate within local political and economic contexts, and this raises questions about their suitability in many political environments. Health workers have been murdered where they have challenged the establishment (Chowdhury 1981; Heggenhougen 1984), and Werner has suggested that while CHW programmes may be community-supportive they may as easily be community-repressive (Werner 1981). These and many other problems are gaining more attention as governments struggle to reproduce on a national scale an idea that has mostly been tested in small areas and in very varied political situations. What remains unchanged, however, is the fundamental dilemma that countries need to improve access to health services with limited resources. This applies equally in urban as well as rural areas. Community health workers therefore remain an attractive policy option for many governments, and urban experience with workers is now accumulating.

While this chapter talks of community health workers, there is considerable variation between schemes. The essential characteristic of CHWs is that they are the most accessible primary health care providers in the community in which they live. They are usually selected by the community, and are expected to be responsive and responsible to that community. They are called many different names, for example, health promoters, friends of health, polyvalent agents, community health workers, and health guides, all subsumed under the name community health worker. In essence, they differ from other primary health care workers, such as nurses, because they are primarily responsible to the community and not the national health services. However, this is by no means always the case. Responsibility largely depends on how they are paid for their services, whether they are part-time or not, and on the range of tasks they are expected to carry out.

Most urban projects appear to prefer CHWs with a broad background, often in nutrition, mother and child health, with organizing and communication skills and an emphasis on prevention and community health education. However, an example of starting with the monopurpose curative approach is the Urban Volunteer Programme in Dhaka, Bangladesh, for promoting the greater use of packets of oral rehydration salts (Stanton *et al.* 1985). This started when the International Centre for Diarrhoea Disease Research, Bangladesh (ICDDR,B), had nearly 100 000 outpatient attendances during 1980 alone. Women volunteers from the poor communities receive a two-week training at ICDDR,B and then return and operate from their own homes. They distribute oral rehydration salts, soap, and Vitamin A for cases of xerophthalmia.

An opposite starting point has been taken for CHWs in Addis Ababa, where they are members of a team of development workers, based within the offices of the local government administration, called the *kebele* (Goyder

Fig. 13.2 A community health worker explains the benefits of basic hygiene to a younger mother in a slum. UNICEF photograph by B.P. Wolff.

1985). However, other programmes have utilized CHWs clearly as health workers, undertaking a mixture of curative and preventive activities, such as Cité Soleil in Haiti (Boulos 1985), the St. John's Ambulance programme in Cape Town, South Africa (Corry 1985), and the primary health care project in the slums of Guayaquil, Ecuador (UNICEF, Quito 1983). Shubert (1986) has provided a useful summary of urban CHWs (see Table 13.1).

Although urban community health workers are a relatively new cadre, they have been, or are, a part of most of the projects outlined in this book. Many of the same issues that surround their rural counterparts have also emerged in urban projects, such as: should they be trained in a wide range of skills or is it preferable to focus their work in a few areas? Should they be concerned with overall development issues or with health activities alone? Should they concentrate on preventive or curative activities? Should they be community or health services based? Since in urban areas many other services are potentially available, the role of the urban CHWs is often less clear

Table 13.1 Comparative chart of urban community health workers

City	Name and method of selection	Role or function	Households covered	Training Time	Training Content	Remarks
Bangkok, Thailand	Village Health Communicator Socio-gram technique	Disseminate and collect health information; mobilize residents for health activities	8–15	1 week	Nutrition, sanitation, preventive health	Peri-urban villages
	Village Health Volunteer Best VHC chosen by Health Centre	Primary treatment, supervise VHCs, manage drug co-operative	100–150	2 weeks + refresher	Simple diagnosis and treatment	Peri-urban villages
	Urban Health Volunteer Selected by Community Committee or Health Centre	Community organization; health, nutrition, sanitation promotion, primary diagnosis and treatment, health information collection, BMN survey	20–30	2 weeks + refresher	Community development, health, nutrition, sanitation, first aid; BMN survey	Congested communities

Colombo, Sri Lanka	Health Warden Selected male/female school leavers by Public Health Department	Community organization, health, nutrition, sanitation education and promotion	100	2 months + OJT	Community development, health education	Paid semi-professional
Seoul, Korea	Community Health Volunteer 'Tong' leaders	Health information collection/ dissemination, referrals	50–100	once per month	Health education and information collection	Paid token amount
Davao, Philippines	Katiwala Voluntary residents usually women selected by community	Simple diagnosis/ treatment, health information collection and dissemination, immunization, MCH, community mobilization	25–30	6 weeks	MCH, treatment of common disease, immunization, promotion of nutrition, sanitation	Both urban and rural areas, incentive payments by residents

Table 13.1 *continued*

City	Name and method of selection	Role or function	Households covered	Training — Time	Training — Content	Remarks
Manila, Philippines	Health Aide Voluntary residents elected or appointed by community committee	Assist Barangay Health Worker on health information collection/ dissemination; promotion of nutrition, sanitation	25	4 days	Preventive health, health information system, medicinal plant cultivation	Great variation in selection, training, etc., not paid
	Barangay Health Worker Previously appointed in nationwide programme	Diagnose and treat simple diseases, referrals, oversee work of health aides	500	not available	not available	Paid semi-professional
Quayaquil, Ecuador	Community Health Worker Women residents 18–35 years with a minimum primary education	Health information collection and dissemination, MCH, immunization, simple diagnosis/ treatment and referral	500	2 weeks + OJT refresher	Preventive medicine, information system	Paid under contract with community

Source: Shubert (1986).
Note: BMN = Basic Minimum Needs
       OJT = On the Job Training

than for their rural counterparts. A crucial issue for urban areas is if and how the CHWs should be paid (Griffith 1983).

CHWs are not necessarily always appropriate in urban areas. An early consideration must be the nature of the urban poor community and its present stage of development (Newell 1985). Different stages can be identified, which are:

- the impermanent and newly settled areas;
- the more permanent and developing neighbourhoods;
- the established and upgraded settled areas.

Antipoverty and social welfare activities are the main concern for the first category. Health measures are perceived as only becoming more important as the basic necessities of employment, food, shelter, warmth, and clothing have been met. It is only the more stable and established communities that give health a higher priority and will use their own resources in this direction (see Table 13.2). Thus, only in the more permanent and established areas are community-based health workers likely to become established, unless they are heavily supported and directed from outside the local community.

These differences have been utilized in Bangkok to organize PHC in phases as outlined in the first case study.

Case study 1   Bangkok Municipal Authority and CHWs

Based on a paper by Sakuntanaga (1985).
In Bangkok, a city of over 5 million people, the overall health resources seem to be adequate but they are mainly provided by the private sector and are not equitably distributed. With the natural population increase and the continuing migration from rural areas, the number of people underserved and lacking basic environmental sanitation facilities is rising. Although there are already 55 fully staffed district health centres,the present health care delivery system fails to meet the basic health needs of the urban poor. People in congested urban areas live near medical and health facilities, but they endure conditions of poverty, poor environmental santitation, and ill health, the same as their rural counterparts. People in peri-urban areas have even more limited access to essential health care.

Initiating Primary Health Care in Bangkok

The Bangkok Municipal Authority (BMA) realizes the inequity of the health situation and considers it to be a high priority. The second BMA Health Development Plan (1982–6) includes the new Primary Health Care Project, which aims to improve health services to the low-income people and to promote collaboration among existing health facilities through a primary health care approach. To achieve this, it is necessary to stimulate, convince,

Table 13.2 Relative importance of different sources of support for health activities at different stages of slum development

Stages of urban development	Priority needs				Sources of support for health activities		
	Anti poverty	Social welfare	Environ-mental	Health care	Public	Community	Individual or family
(1) New migrant areas	+ + +	+ + +	+ +	+ +	+ +	+ +	+
(2) Developing neighbourhoods	+ +	+ +	+ + +	+ + +	+ + +	+ + +	+ +
(3) Established suburbs	+	+	+ +	+ + +	+ +	+ +	+ + +

Key   + least important
          + + fairly important
          + + + most important
Source: adapted from Newell (1985).

and support communities and individuals to help themselves; to develop appropriate attitudes and skills among all levels of health personnel, especially those working in the community; and to involve the people in health programme planning. Priority is given to three types of poor and disadvantaged communities, which are:

1. Peri-urban villages: low density, traditional rural community organization, difficult access to health services, low level of urban infrastructure, mainly agricultural economy.
2. Unimproved congested communities: high density, no recognized community development organization, no legal claim to land, low level of infrastructure, close to health facilities, mostly informal sector employment.
3. Improved congested communities: high density, recently organized Community Development Committees, legalized land tenure, some infrastructure improvement, close to health facilities, mostly informal sector employment.

priority comm^s

In peri-urban villages, the well-established national PHC programme methods were used with only minor modifications. Village Health Communicators (VHC) were selected and trained to disseminate health information and become the contact point for their group. VHCs are trained to:

- analyse and list priority health problems of their own community;
- co-operate in planning and other health activities;
- communicate basic knowledge on health and sanitation to family members in their groups; and
- inform health officials about those problems they cannot tackle by themselves.

VHC

After serving their communities for 12 months, one Village Health Volunteer (VHV) is selected from among each group of 10 VHCs. Additional training in simple diagnosis, treatment, referral, and management of patients is provided and the VHVs are given basic medical kits to treat minor ailments. VHVs also set up and manage the drug co-operative for the village. They refer serious cases to district health centres and hospitals, and organize immunization and other campaigns.

VHV

The improved congested community is an organized community in which the infrastructure has been improved by the National Housing Authority or BMA, which have helped to establish committees elected by household heads in that community and officially appointed for a period of 2 years. In these communities, another type of volunteer—Urban Health Volunteer (UHV) has been developed. These volunteers are chosen by the community committee and approved by the district health centre on a basis of one per 20–30 families. Some of them will be members of the community committee. They are given training and tasks similar to the VHVs. However, they may be given additional training in the areas of environmental sanitation and drug abuse. The UHVs have to become involved in general problems of community development, to which health is closely interrelated in the urban areas.

UHV

The BMA has only recently begun work in the unimproved congested areas, where several serious constraints have been encountered. They are not legally recognized residential areas and they have no recognizable community organization. As a result, many government services cannot be legally provided in these areas and infrastructure and housing improvement is restricted by the landowners.

Although base line surveys have only recently been completed, the BMA has already begun to work with three selected unimproved congested communities. Experience has shown that the UHVs in these communities must be given special training to function as community organizers. The health problems and PHC activities are dependent on the creation of an effective community organization and on other services that are often not presently available.

The VHCs, VHVs and new UHVs are all supported by well-trained and well-equipped medical teams in the 55 District Health Centres. Each centre has one or more doctors, nurses, midwives, public health nurses, and usually a dentist and social worker. The public health nurses frequently visit these communities to monitor and guide the work of the volunteers.

Constraints

In Bangkok 1.2 million, or nearly 25 per cent, of the population are housed in congested or squatter settlements, many renting land and building their own houses. About 63 per cent are on private land. Land rent is low and landowners often arrange to supply water and electricity in the houses. The environmental conditions, however, are very poor. The land is marshy and has no drainage. The lack of tenure makes people reluctant to improve their houses and landowners discourage permanent construction. The BMA has no authority to make physical improvements in these areas without the permission of the owner. Recently there were massive land purchases on the urban fringe which limits expansion of low-income areas and increases pressure on low-income communities inside Bangkok. This tenure insecurity makes the people feel more uncertainty and discourages community participation.

Informal sector occupations and economic pressure may impose limitations on the performance of volunteers and self-help activities, especially in the areas that are dominated by temporary labourers. However, the areas inhabited by the lower middle class, often with permanent jobs, have thriving community organizations and a good potential to cope with their own problems.

Urban medical sophistication and the proximity of drugstores and other modern services limits the importance of UHVs in dispensing drugs and, therefore, curative care does not give them as much creditability as it does in rural areas.

Health managers are a critical factor in the success or failure of PHC. It requires tact and diplomacy to introduce such innovations. Health services must be interlinked from the grass-roots level to high level at health centres and hospitals by means of patient referral. The health infrastructure needs to be strengthened and reorganized to link different health programmes into one unified system. Health services in Bangkok are not only organized by the BMA but also by other government and private institutions.

Many health-related services are outside the control of BMA, such as water supply, refuse and sewage disposal. Physical improvements are often carried out by the independent National Housing Authority (NHA). Furthermore, there are a large number of NGOs working independently in congested communities. Recently, meetings and workshops have been held among these agencies and NGOs, but systematic co-ordination is still lacking.

Finally, health personnel themselves have to change their role and attitude from one of sitting in the office and waiting for the patients, to one of going out and working with people in the community. They must help people to help themselves. The great availability of hospitals and health facilities has encouraged health workers to think that the use of UHVs will not work, and many even question the need of UHVs in the existing health system. PHC in the city is a new movement, and implementing it is a difficult task, but it must be done.

Case study 2 The Katiwala Programme, in the Philippines

Based on a paper by de la Paz (1985).
The Katiwala programme began as a free medical clinic in 1967, supported by the Christian Family Movement, to give health services to the residents of squatter areas in Bajada district of Davao City. The clinic was supported by regular donations from commercial establishments and private citizens and it was staffed by volunteer health personnel.

After two years the clinic staff realized that there was no improvement in the health of the local residents and that the free clinic was detrimental to human dignity. The clinic was closed and a series of meetings were held between the clinic staff, the families, and a professional social worker, which resulted in the transformation of the free clinic into a medical co-operative. The members agreed to pay minimal fees and to buy the medicines prescribed at wholesale or subsidized prices. They assumed some duties in the management of the clinic, such as helping the health personnel and maintaining order among the patients. They started to have a voice in its policy making and in the day-to-day operations.

General meetings of member families were held every 3 months or when the need arose. Eventually, area leaders were chosen by the families to represent them at the meetings and they served as a liaison between the members and the clinic staff. The need to raise incomes became clear at one of these meetings, which led to the establishment of a sewing workshop. Volunteers initially trained the local housewives in sewing and later in quilting and bag making.

In 1972, after lengthy consultations between the community members, the staff, and the donors, the Development of the People's Foundation (DPF), was formed as a non-profit foundation. Its main purpose was to manage the medical co-operative and the sewing workshop. Although the medical co-operative consisted of only 500 families from 31 low-income communities near the clinic, the volunteer health staff could not cope with the number of patients who came on clinic days, held on two afternoons each week. At a subsequent general meeting, the members and the staff decided to undertake the first volunteer community health worker (VHW) training. Since the workers would be living in the same communities, the VHW would bring the services closer to the people who most needed them—people who did not have enough money for fares or who had no one to leave at home to watch over their children and their possessions. The volunteer health workers, who called themselves *Kauna-unahang Katiwala ng Kalusugan* (Primary Trustee of Health), and later shortened to Katiwala, rendered simple curative services to their members in their homes. The first group of Katiwala were trained by the DPF in 1972.

A total of 167 Katiwala were trained from 1972–8, but only 97 were from the 31 low-income areas. The rest were sponsored by non-government organizations for services outside of Davao City.

Institute of Primary Health Care (IPHC)

In 1977, the Davao Medical School Foundation was formed as a consortium of well-established institutions, including the DPF. The founders decided that training Katiwala on the same campus as the medical students would promote opportunities for teamwork, stress primary health care, and promote a spirit of service to underserved people. Subsequently, the Institute of Primary Health Care (IPHC) was established to implement the Katiwala programme.

The objectives of the programme were:

- to render primary health care to the low-income residents in urban slums and underserved rural areas;
- to mobilize the community to participate in PHC;
- to promote co-ordination and co-operation between the community, the health system, other government agencies, and private voluntary agencies in the formulation and implementation of PHC activities;

- to build a strong and lasting community support system for the Katiwala so that together they could work for their own socio-economic development.

In 1978, it was agreed that the DPF would continue to serve the urban areas and IPHC mainly the rural areas. From 1981 both organizations have worked together in the same urban areas, with Katiwala activities serving as entry point into the *barangay* for IPHC.

Selection and recruitment of Katiwala trainees

By 1975, the DPF employed a full-time project co-ordinator who also served as training officer and community organizer, conducting home visits, family interviews, small group meetings and assemblies. He was assisted by volunteers from the community in conducting a baseline survey to identify health, nutritional, environmental, and socio-economic problems, and to make a list of the possible Katiwala activities. Selection of the Katiwala trainee was done at a general assembly and their work during the surveys tested their suitability for further training.

Similarly, the IPHC employed general assemblies, held after a three-month community preparation period, to select the trainee. Later this was extended to six months.

The criteria for selection were:

- resident of the community;
- willing to be trained and to serve;
- credibility, and acceptable to the community;
- functional literacy;
- physically fit;
- time to serve the community;
- above 20 years of age and married;
- sex—no preference but candidates mostly women.

Training

For DPF this started as one day per week for six months but is now daily for one month with practical work on clinic days. However, since IPHC is mainly training rural Katiwala the course schedule is two weeks living in, with two weeks' practical in their own villages, and a further two weeks living in. Whereas the IPHC uses the Davao Medical School Foundation (DMSF) buildings, the rural Katiwala are trained at a centre near their own villages.

The early training by DPF was decided on by the trainees themselves, based on their observations and experience in the community. In 1975 a training officer was appointed and the training was oriented to the tasks the Katiwala performs in the community and the clinic, so that the training became more competency based and task oriented. It was conducted in the local language.

The community survey is collated and analysed by the Katiwala trainees, and group work focuses on community needs. Visual aids and teaching materials have been prepared to suit local conditions. Small group discussions, brain storming, and role-play are used whenever applicable. Flow charts have proved popular. The training materials are in a loose-leaf folder and draw heavily for many topics from *Where there is no doctor* by David Werner.

The training staff were volunteer physicians and nurses who were sympathetic to the training approach, but as the project staff became more confident in using participatory teaching methods, reliance on visiting professionals decreased.

The course covers:

- a sharing of experiences and observations during the community survey;
- community organization and self-help;
- human relations and communications;
- analysis of community health situations and health delivery;
- prevention and treatment of common conditions such as cough, fever, diarrhoea, injuries;
- prevention and treatment of immunizable diseases;
- nutritional deficiencies and promotion of better nutrition;
- maternal and child health, including family planning;
- environmental sanitation, personal hygiene, and parasitic infections;
- herbal medicines;
- teaching skills, preparation of visual aids;
- working with agencies in the community;
- planning and preparation of Katiwala activities;
- vegetable gardening is included for rural areas.

Practical skills are gained in the community and in the health clinics under supervision. Although the core curriculum remains the same, changes are made or topics are added as necessary. The same teaching techniques are later employed by the Katiwala when conducting family health classes.

Continuing education

Meetings of all the Katiwala take place at which they discuss their performance, problems they encountered and how they solved them. By sharing they realize that other Katiwala have faced similar problems. Eventually, continuing education is done by the Katiwala themselves with the IPHC staff playing a minor supportive role.

The Katiwala Action Plan

This is an innovation introduced by IPHC. Before the end of the basic training, *barangay* officials and representatives of service agencies working in the *barangays* join the Katiwala and together they draw up a plan of activities, based on the community survey. This plan, the Katiwala Action Plan (KAP) enables the villagers to assume more responsibilities for the health activities of the Katiwala and prepares them for the eventual phasing-out of the IPHC staff. Upon her return, the Katiwala presents the plan to the community, and together they agree on how best to implement it. When the Katiwala plan calls for activities already being done by other agencies, the Katiwala co-operates with them, as in the nutrition surveillance done by the nutrition workers or the nutrition education classes conducted by the extension workers of the Ministry of Agriculture. She assists the government health personnel in campaigns for immunization by informing the mothers in her *barangay* about the schedule, by gathering them on the specific date, and helping keep order during the actual immunization days.

Katiwala activities and responsibilities

The Katiwala is responsible for 30–50 families in her *barangay* and experience shows that she devotes 1–2 hours, two or three times a week to her voluntary work.

The Katiwala have identified the services they render in the community as follows:

- *barangay* activities such as social gatherings,
- linkage with government and non-government organizations,
- implementation of the Katiwala Action Plan,
- provision of Primary Health Care Services.

These health activities include:

- home visits to treat simple complaints like cough and fever, give first aid or advice on health and nutrition;
- campaigning for environmental sanitation;
- organizing and conducting family health classes;
- referrals to health centres;
- growth monitoring;
- attending maternal and child health care clinics;
- keeping of records;
- links with representatives of other health agencies.

Credibility and acceptance of the Katiwala is enhanced by the simple curative skills she has acquired. However, over the years the Katiwala, although primarily a health worker, has gradually been drawn more and more into general development activities.

Community support systems

The Katiwala support system has gradually been strengthened. The first step is taken by the IPHC staff when they approach the local officials for permission to discuss the programme with the villagers. Following discussions, a full village assembly is convened where the role of the Katiwala, the community, and the IPHC are explained. Besides home visits, small group meetings, and assemblies, IPHC has tested other strategies for generating community participation and involvement, but feedback from Katiwala already trained indicated that community support was sporadic and short lived. One strategy used in 1981–2 was the Community Leader's Training for local officials, church leaders, members of women's and youth groups, and other interested residents. Topics discussed included the Katiwala programme role clarification, health situation analysis, leadership training, problem solving, and communication skills.

The Community Leader's Training was meant to form a core group that would assist in disseminating health information, selection of the Katiwala and help her in the performance of her activities in the *barangay*. These core groups were able to help in the selection of the trainee and were active in assisting the Katiwala immediately after her graduation. They helped present the Katiwala Action Plan to the community. Gradually, though, these groups lapsed into inactivity. Was it due to a lack of motivation from IPHC or a lack of a clear function as a support group?

The current strategy for stimulating community involvement is the Team-Building Workshop. Meetings are held at provincial, municipal, and *barangay* levels with representatives from all government agencies, *barangay* officials, and identified *barangay* leaders, including the Katiwala. An action plan utilizing the resources of the agencies represented is drawn up. *Barangay* meetings follow the municipal meetings, attended by *barangay* officials, formal and non-formal leaders, and representatives of agencies who have made a commitment at the municipal level. The Katiwala is one of the leaders who attends the municipal and *barangay* sessions, where she brings the health problems to the attention of the other leaders and obtains their support. Although the initiative comes from IPHC, the follow-up comes from the *barangay*.

However, in the urban slum area of Agdao where the Katiwala have been active for almost 10 years, community involvement was not sustained. For these communities the IPHC devised a new strategy. Focused Family Dialogues (FFD) were held where individual families discussed their aspirations in life, the obstacles, and the possible solutions. After these dialogues, the whole community was gathered in a Focused Community Dialogue (FCD) to discuss the individual aspirations. Each family was then asked who they could work with harmoniously in order to form small working groups or clusters. The clusters further discussed their plans and activities and devised schemes for implementing their ideas. The obstacles identified are often economic and the solutions are activities to generate capital or income, or to obtain small loans to finance small businesses. The IPHC realizes that health problems cannot be dissociated from the other realities of daily life and that once the community learns how to cope with its economic difficulties it can attend to its health needs. For this reason, IPHC has become more involved with communities in economic activities. At present the clusters are doing well. Many have engaged in small businesses, like selling firewood, repacking commodities for sale, sewing and setting up small stores. The clusters are active, not only in their income-generating activities but also in assisting the Katiwala perform her tasks.

Health worker support

In general, there is rapport between the Ministry of Health personnel and the Katiwala. The midwife or nurse in the health center provides technical supervision, just as the community support group supervises her community activities. When necessary, the Katiwala brings villagers to the health center for referral. The midwife or nurse in turn enlists the help of the Katiwala for informing the villagers regarding schedules for immunization or weighing of 0–6 year-olds.

A unique support group for the Katiwala are the Health Scouts. These are school children from 8–15 years of age who have been trained to help their pre-school siblings in their physical, mental, and spiritual development. Trainers are called Child Trainers—volunteer workers who have received special training from IPHC. The Health Scouts are taught how to prepare ORS, how to use the growth chart, how to promote the benefits of personal hygiene and environmental sanitation, and nutritional education. The Katiwala is their adviser and together they plan their activities. The Health Scouts thus help the Katiwala in various other ways, such as information dissemination and mobilizing the community.

Incentives and payments

The Katiwala is a volunteer and she does not demand payment for services rendered, but she is free to accept tokens of gratitude given in cash or in kind. It is customary for the family to give fruits, other foodstuffs or money, yet many more can only say 'thank you'. The community realizes that the Katiwala is in the same financial situation as the families she serves. The residents have tried raising funds on an *ad hoc* basis through:

- benefit dances, raffles and bingo socials;
- allocating *barangay* funds;
- using profits from the village drugstore;
- income-generating groups to which the Katiwala can join;
- various non-monetary incentives also motivate her to continue her volunteer work, such as attendance at continuing education meetings, annual conventions, and attendance at seminars on primary health care.

A newsletter in the vernacular, called '*Lanog*' and a weekly radio programme serve as a link between the Katiwala and the IPHC.

The role of local and international organizations: scaling up

From an institution engaged in training Katiwala for underserved areas in
South Eastern Mindanao, IPHC has expanded its scope to other islands and
other provinces of the Philippines, sharing its philosophy, its training
methodology, its training materials, and transferring its technology to other
training teams. This would not have been possible without the assistance of
many different government and non-government organizations.

The IPHC strategy is presently being used on a regional scale and there is a
commission for its national implementation. In addition, IPHC has trained
volunteer health workers for Slum Improvement and Rehabilitation areas
(SIR) in Metro Manila, Tabago, Albay, and Cebu City.

A local Interagency Advisory Council was formed in 1978, and a formal
support agreement was signed by the Ministry of Health and IPHC. The
Council's main function is to support PHC. The Regional Development
Council adopted the Katiwala approach in 1979 as a component of the
region's development strategy. The council is supplemented by Interagency
Committees at provincial, municipal, and *barangay* (village) levels. Members participate in the Katiwala Action Plan (KAP) and the MOH.

IPHC, assisted by the Asian Health Institute (AHI), has conducted
seminars in planning, implementation, monitoring, and evaluation of
community development projects for intermediate level workers from the
Philippines and South-East Asian countries. AHI has contributed to staff
strengthening through study grants awarded to the IPHC staff and through
assistance for the IPHC Project annual review.

The association with UNICEF began in 1978 when IPHC linked up with
the government and held workshops to engage experts in the field of community organization, participatory teaching methods, planning, implementation, monitoring, and evaluation. The expansion of IPHC activities
from health to a development approach was supported by UNICEF. Requests
for training continue. IPHC now has several training teams each with its own
area of competence—primary health care, child to child training, farmer
training, community credit, small businesses. At the *barangay* level, the
Katiwala, the credit group officers, and the farmers' leaders have already
started training neighbouring *barangays*. Thus, health and Katiwala activities
have led IPHC into the necessity of supporting general development and not
just that of the health sector alone.

Comments and conclusions

- Only in stable and reasonably well-established urban communities, where community structures have developed and many basic needs are being met, are urban CHWs and their health activities likely to receive local commitment and support. Medical care for acute illnesses is the major priority in less well-established slum areas, rather than preventive activities.

- Just as with rural CHWs, those in urban areas need to be an integral part of their local community, and selected and supported by them. However, outside skills, financial support, and supervision by health services staff continue to be crucial to their success.
- Unco-ordinated and different policies followed by different agencies, government and non-government, are a major problem that undermines progress towards community based and supported health activities.
- Financial support for CHWs, whether salary- or income-generating activities, is a more crucially important factor in urban communities where the cash economy prevails.
- Health activities tend to focus on mothers and children, nutrition, immunization, oral rehydration, water, and sanitation.
- Many urban CHW programmes have had to develop a multi-sectoral approach and become involved in such activities as *crèches*, schools, housing, and income-generating activities.
- It is easy to start small-scale urban CHW programmes, but at least 10 years are necessary before they become well established and well integrated into the local community and accepted by the local municipal organizations.

14

Scaling up and scaling down

Going to scale and the scaling up of pilot projects or of limited efforts to reach a national coverage with broad, maximum participation, present the most intricate and intractable problems in development. Literature on rural and urban development abounds with examples of projects that could not be expanded in scope or coverage to reach a larger target population. However, there are also successful operations that have achieved national coverage and have been able to sustain their achievements—some examples being the Green Revolution in the economic sphere, and literacy campaigns and movements in the social sector . . .

Nyi Nyi, 1984, p. 25

. . . there still remain large groups of people—particularly the poor—who lack health care coverage and access. As yet the health system as a whole shows little sign of absorbing the message and making the fundamental adjustments required. For no health system can be effective that lacks a comprehensive, low cost service of first contact, or fails to use its more sophisticated and expensive resources in support of this primary health care network . . .

WHO, 1984b

The two processes

The above quotes point to the problem of scaling up pilot projects in general and more specifically to the obstacles that can stand in the way vf extending primary health care coverage. Throughout this book successful pilot projects have been identified, but few of these have yet been expanded or replicated on a significantly larger scale. On the other hand, centrally planned large-scale campaigns, such as those for immunization and family planning may successfully reach a significant proportion of the poor, but 'tend to have a high profile and are short-lived, they therefore rarely maintain their momentum in the long run . . .' (Nyi Nyi 1984). In the case of primary health care (PHC), the problem of scaling up is far from academic. The very concept of PHC grew out of the urgent concern to find a way of assuring 'Health For All by the Year 2000'.

This chapter is concerned with two processes: *scaling up* primary health care projects from the local level to the city, provincial, or national level, and *scaling down* health programmes from the national, provincial, or city level

194

to the local level. Very often both elements are needed simultaneously. There are two key questions which are complementary:

- How can local projects be expanded or replicated to achieve universal health coverage of the population in a given area? (Bosnjak 1985.)
- How can top-heavy health services with top-down planning methods, norms, and procedures be adapted flexibly to serve the specific needs and characteristics of local groups? (Tabibzadeh 1985.)

Scaling up and scaling down mean, in one sense, that policy makers and administrators at the top levels of government need to undertake more collaborative health planning in order to design or re-design national programmes, so that they are adapted to, and responsive to diverse local conditions and communities. At present programmes are often handed down as standardized, pre-designed, 'pre-packaged' programmes, to be implemented in a uniform manner across the nation, province, or city. It also means thinking about 'scaling up' projects which have been effective at the local level to the system-wide level—nation, state, city—through expansion or replication.

Constraints to scaling up and scaling down

The Report of the Joint UNICEF/WHO Consultation on Primary Health Care in Urban Areas held in Guayaquil, Ecuador, in 1984, neatly summarizes some of the main constraints which stand in the way of both scaling up and scaling down (WHO 1984*b*).

- The input of financial and human resources in pilot projects is often too high to be widely replicable.
- Any single 'standard package' may not have enough flexibility within it to be adapted to a wide variety of specific local situations and cultural contexts.
- Often voluntary organizations have initiated projects with no or little involvement of central or local government; thus, they have not had to deal with the real constraints of government bureaucracies and therefore they have a low credibility with government.
- Often there is insufficient community involvement at all stages to assure appropriateness, cost effectiveness, coverage, and continuity.
- The innovative project leaders or government officials may be in too much of a hurry to establish a large programme and to go to a larger scale.
- Attitudes of government officials and the health professionals are sometimes major constraints.
- Government officials are also often sceptical or suspicious of non-governmental agencies.
- Medical and other health professionals are usually suspicious and resist such radically new approaches as PHC.

Fig. 14.1 Scaling up actions in the urban environment often means reaching physically isolated communities, including homes like these in Mexico at the top of the highest hill. Photograph by Vogler, OXFAM.

Lessons from success

On a more positive note, there are in Myers' (1984) words some 'lessons of success'; i.e. important elements that can be identified in successful expansion from local projects to large-scale programmes. After analysing the studies of Korten (1980), Paul (1982), and Pyle (1984), Myers himself identifies five issues as being fundamental. They are:

- What are the financial costs of going to scale and who bears them? The community, the government, international aid agencies?
- Successful larger scale programmes seem to begin by offering a single service. Multiple strategies can easily overload the organization while the concept of 'converging services' seems to be more appropriate. That is, different organizations may deliver single services at the same time and place. While community participation is necessary, continuous participation is difficult to achieve on a large scale; and such participation is probably impossible without recruiting staff or technicians sensitive to community needs and problems.
- The role of communication is important. Mass media can be used to mobilize demand, and social marketing can be used at the highest levels of government as well as at the community level.
- Evaluation should be flexible, continuous, and participatory from the outset.

There is need for a new type of evaluator who can work with local people and local staff, and a need to provide quick feedback of results to both communities and government.

- The role of international assistance agencies should include:
  - (a) support for relatively expensive pilot projects with the understanding that few will ever be capable of going to scale but that valuable lessons can be learned from them;
  - (b) a willingness to use flexible management methods which can take advantage of opportunities;
  - (c) supporting committed individuals who appreciate the processes involved in scaling up yet may be tied to a small-scale project;
  - (d) the acknowledgement that different agencies can play different roles (for example, small voluntary organizations might support local level projects and larger agencies could support intermediate-sized projects/ programmes in the scaling up process).

Development activities in fields such as environmental sanitation, drinking water, education and, above all, employment and income generating have a profound effect upon health. However, programmes which are unco-ordinated with other sectors will not only bring about minimal effects on the improvement in the health status of the population, but will also increase overall expenditure; since each programme will require its own resources in staff, administration, and infrastructure. Co-ordinating health, health-related, and other social welfare programmes not only has a synergistic effect but also generates resources from all the different sectors which can be combined for the benefit of the people. This is the reason that Bosnjak (1985) advocates 'an active search for new partners' at the local level. At the highest levels of government, such co-ordination requires major policy decisions (Tabibzadeh 1985).

Learning while doing: information-sharing, monitoring, and evaluation

Learning from existing information is both a vertical and a horizontal process and it appears to be essential for successful scaling up. It is vertical in the sense that planners and policy makers learn from local experience and vice versa. It is horizontal in the sense that there can be information-sharing among projects with similar objectives. Bosnjak points out that qualitative information from local projects 'can enrich decision maker's understanding of quantitative data from secondary sources . . .' and can enhance their understanding of poor people's conditions and felt needs. There is a need for simple information systems which permit local-level workers to gather data for evaluation. Such systems are rare, but appear to be a necessary and essential condition for effective learning from experience. This is one of Korten's conclusions. He says, 'The learning process approach calls for organizations that: (a) embrace error; (b) plan with the people; and (c) link knowledge building with action.' (Quoted in Myers 1984.)

Case study 1   Reaching out into the *kebeles* of Addis, Ethiopia

Based on a paper by Teferra (1985).
'Pilot projects are not too good to be replicated' Jember Teferra.

The initial problem in scaling up that most projects will face will be extending activities to other communities or neighbourhoods. As an example of a programme which is attempting to scale up its activities at this limited level it is interesting to focus upon one which has been presented earlier—the integrated community development programme in Kebele 41, Addis (see Chapter 11).

Plans have been implemented to scale up the activities in Kebele 41 to reach into five neighbouring *kebeles* with a total population of 35 531 (Teferra 1985).

Socio-economically, all five *kebeles* have certain things in common, particularly all have very poor people living in them who are unable to adequately support themselves.

From the health point of view the water and sanitation problems in the neighbouring *kebeles* are almost identical to those in Kebele 41. Some of the *kebeles* have the added disadvantage of being totally unreachable by road. The only distinct difference noticed was the eagerness to benefit from existing health facilities, because seeing is believing and they have seen changes in their neighbouring Kebele within the past four years. The usual comment is 'Kebele 41 which was the bottom of the scale is already at the top now'.

Activities already scaled up

- Water. This activity was extended to the neighbouring *kebeles* at an early stage as the two municipal water taps which existed in 1981 were inadequate, with people queuing all day. This was solved when *Redd Barna* put up one water post with four taps and another two with six taps each. Until the community needs increase this is adequate.
- Immunization. The outreach programme to the neighbouring *kebeles* was started as early as 1983. Once the Kebele 41 under-5s were covered, the neighbouring *kebele* administration and health committees were contacted and all the five *kebele*'s children under 5 years old are being vaccinated.
- Health education. This began at the same time as the vaccination outreach was started. Prior to starting the immunization programme the same methods as those used in Kebele 41 were applied. Film shows and discussions go on.
- First aid. Also started in 1983 at the same time as work in Kebele 41 began. People needing first aid assistance or an ambulance from the Red Cross for emergencies are assisted through the youth volunteers during the night and education extension workers during the day.
- Sanitation. The sanitary guards and CHWs together with health committees of

the neighbouring *kebeles* have been organizing clean-up campaigns periodically. The sanitary guards became more involved in 1984 when they co-
ordinated with the health committees in each *kebele*.

- Shower services. The neighbouring *kebeles* are benefiting from the cold shower
  in Kebele 41, paying ten cents (US) a time as often as they want to wash. Even
  other *kebeles* beyond the five neighbours are allowed to use the facility, though
  priorities are given to Kebele 41 and the neighbouring *kebeles* if there is a water
  shortage or too great a demand.
- Family planning. Registration and education in neighbouring *kebeles* has also
  been introduced. The number of volunteers is encouraging.

Problems encountered in scaling up

- Non-integration of activities. The Kebele 41 project emphasizes integration
  and a holistic approach as the only way to succeed in achieving Health For All
  by the Year 2000. The scaling up and enlarging of programmes, or outreach,
  that is being done from Kebele 41 to the five neighbouring *kebeles* is just part of
  the integrated programme. Other elements, such as housing improvements
  need to be scaled up. Teferra (1985) asks 'What would be the point of teaching
  a household with an over-flooded latrine to keep the environment clean or
  cover their toilet when they do not have the financial means to do so? Or what
  point is there to teach the community to collect their rubbish in containers and
  dump them into trucks when they don't have the bins?'
- Lack of full-time health workers. As an example of the problem, the follow up
  of children for the completion of immunization programmes is very tedious
  work which took both the CHWs in Kebele 41 a year of constant community
  education, house-to-house visiting, and looking for defaulters before
  achieving the current record and immunization programme. More CHWs in
  each *kebele* are needed.
- Poor co-ordination between *kebeles*. There are two *kefetegnas* (higher
  administrative units) in these neighbouring areas. When the Kebele 41
  chairperson arranged meetings in order to upgrade the health committees of
  the six *kebeles* involved, the *kefetegna* representatives failed to attend. Being
  unable to make a financial commitment, *Redd Barna* staff found it difficult to
  call an official meeting which might raise expectations. But, even with adequate funds, co-ordination will be a problem that has to be settled in the early
  stages by having one full-time co-ordinator to liaise not only with the different
  kebeles but also with other agencies.
- Weak supervision. The existing Kebele 41 staff could not supervise neighbouring *kebeles* to the same standard as in Kebele 41. From the Kebele 41 experience
  the supervision must be from the community and this takes time to establish.
  More staff are needed to mobilize the respective communities.
- Lack of funds. To scale up the remaining health programmes and some community development programmes (e.g. income generation) a budget was
  worked out for all five *kebeles* to cover a period of three years. It came to Birr
  2 600 000 (or US$1 300 000). The improvement of *kebele* halls, building of

latrines, and the cost of latrine suction trucks have been included in the budget. It will be difficult to raise such an amount.
- 'The problems are there and the solutions are there. It remains for interested organizations to take up the challenge and tackle the work.' (Teferra 1985.)

Some of the problems outlined above also occur during large-scale expansion. After scaling up to cover the entire city comes the challenge of repeating such activities in other cities. The next case deals with this larger-scale expansion, which is, to date, very unusual.

Case Study 2 'Growing' to scale with Indian cities

In August 1984 the Government of India and UNICEF together reviewed the Urban Basic Services programme that was then being implemented in 42 towns (see UNICEF 1984*b*, 1986). The UNICEF Urban Basic Services programme emphasizes the following priority areas: malnutrition, women's development activities; pre-school, day care and early childhood development; responsible parenthood and family planning services; support for abandoned and disabled children; and water and sanitation. Following a community-based approach, the programme involves government and non-governmental organizations and encourages the incorporation of the above activities in government physical development projects.

The Government of India (the Ministry of Urban Development) decided to expand the Urban Basic Services (UBS) programme to 295 towns during the Seventh Plan period, i.e. 1985–9. The Ministry agreed to share programme costs with the State Governments, local bodies, and UNICEF. This not only manifests the commitment of the Central Government towards the programme but is a source of strength and encouragement to poor state and local governments. Fonseka (1985) has described the administrative and operational changes needed in order to facilitate this rapid expansion. Changes included the larger administrative unit of the district becoming the planning unit (rather than individual towns).

Thus, the Urban Basic Services programme in India is expanding fast. It is growing and now reaching a large proportion of urban poor children and women. In Fonseka's words: 'perhaps it is *growing* to a scale'. Fonseka suggests that this Indian experience pin-points a minimum number of 'action steps' that are necessary to create the conditions for expansion. They are:

- a long phase is needed to demonstrate that the project works well,
- it is the people's satisfaction that makes the project 'the talk of the town',
- mobilization of the media is needed in order to communicate the experiences of the project,
- an early incorporation of the project into the city's plans is helpful,
- a cautious period of expansion to an additional few cities is needed to prove replicability and sustainability.

This demonstration phase takes a minimum of 4–5 years to achieve 'programme-status' and public confidence. Fonseka emphasizes the need for a greater awareness and recognition among communities and officials of the essential role that the former need to play at the project-level reviews. In Colombo, Sri Lanka, for example, the community leaders nominate six from among them to be members of the Project Management Committee headed by the City Mayor (Peries 1985).

Comments and conclusions

Similar problems to those experienced in the above case studies have occurred in an attempt to scale up activities in Manila. Here the reorientation of government urban health services, which was initiated in 1982 in four low-income areas (*barangays*) of the city (the Pandacan area) as a research

Fig. 14.2 The problems of scale often seem insurmountable. *Source*: D. Brunner.

and development initiative, has been gradually extended to 18 other low-income areas and is shortly heading towards total city coverage. Further-more, the Manila experience in urban PHC is now being used as the basis of health development in other urban areas throughout the country (see Suva 1984). Constraints and problems have been encountered in funding, *intra*-sectoral co-ordination (*sic*), political factors, and frequent drop-outs among trained *barangay* health volunteers (as opposed to the *barangay* health workers, who receive payments). So, the problems experienced in scaling up are often similar in different situations.

However, the conventional analysis of existing projects and preparation of case studies often fails to shed light on such problems and do not produce information which is needed for scaling up initiatives. Bosnjak (1985) suggests that in future the following analytical framework should be used in a systematic way:

- focus the description of projects upon replicable elements which can be gener-alized and not the ones which were unique in a particular situation;
- analyse the cost-effectiveness and the social acceptability of the project activities;
- analyse and learn from both the achievements and the failures;
- note the relevance of qualitative and quantitative data for stimulating action;
- clarify the advantages of the people's participation and how this was achieved.

15

In the shadow—policy issues

The plight of those living in the slums and shanty towns of the developing world is one of the most important, yet least understood, problems facing the human race. Many worthwhile projects are now in place, and there is an increasing fund of experience on which to draw. More people need to be aware of the scale and urgency of the problem, and of the approaches that should prove relevant, particularly those in positions of leadership in the cities and countries of the developing world and in international organizations. Above all, more action is needed and rapid scaling-up of current efforts from isolated initiatives, until primary health care for people at risk in urban areas becomes a central component of national and international health care strategies.

WHO, 1983*b*.

Identifying and understanding poor urban communities

The need to identify the poorest of the poor and reach beyond sophisticated political structures and self-interested 'leaders' was expressed in many of the

Fig. 15.1 The adverse health conditions of the urban poor are often hidden away. More light needs to be shed upon them. Photograph by V. Virkud.

case studies presented. But before one reaches this group a community needs to be identified. Where do you find the community in the fragmented and uprooted populations of the slums? Sometimes the slums are too unstable for any form of community cohesion and structure to arise. Certainly the heterogeneous nature of poor urban communities is one of the main differences between the rural and urban poor. However, these case studies have indicated that new kinds of social dynamics are operating amongst urban people, arising from old social patterns and modified by the new realities of inter-dependence and a struggle for survival. The sharing of common experiences and involvement in each other's problems begins to forge a new community with its ties of friendship and obligation (Kingma 1977).

Organizing communities by clusters or geographical boundaries may be easier to do in the relatively more homogeneous rural communities. In urban areas, due to the heterogeneity of the slums, the communities may need to be organized around common groupings, such as women's groups, employment opportunities, or on religious beliefs.

The joining together of these different groups eventually may enable larger community organizations to be activated. Such a strategy definitely enhances people's participation.

Once the community has been defined, implementing a process of community diagnosis can help to identify the poorest of the poor and the community's felt needs. This approach has proved useful in many of the programmes discussed. We know that in poor urban areas there may be a bigger gap between needs and demands than in rural areas. For example, with the well established reliance on and demand for private practitioners, hospitals, pharmacists, and other curative services in urban areas, it may be difficult for a community to perceive preventive health care as a need. Often preventive health services are perceived as second-rate in this context. One of the aims of community health education is to make explicit the unfelt needs and convert them to felt needs (see Fig. 15.2).

Despite this established reliance on curative health care, the urban poor often have a clear social interpretation of causes of ill health and they often recognize poverty as the main cause. In this sense the urban poor's interpretation of health problems may differ from that of the rural poor. Work in Peru, comparing rural and urban primary health care, found that some rural communities had a traditional interpretation of health and blamed their health problems on individual failure to make suitable offerings to the gods (Muller 1983). In this context there is little motivation to organize a collective struggle against disease. A purely biological interpretation of health, which leaves the struggle against disease to the health 'experts', is also inadequate. This is perhaps the interpretation that is expected of poor urban communities who have the aforementioned reliance on curative health care. This is certainly true in some situations but in many of the case studies in this book

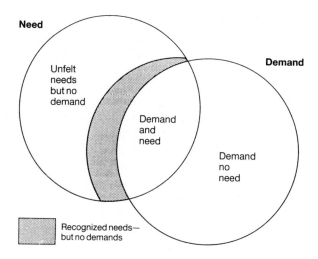

Fig. 15.2 Demand and need. *Source*: Bennett (1979).

we have seen communities taking a much broader, social and political interpretation of health.

Health planners, administrators, and professionals may not understand
the underlying causes of ill health in the same way. As Chambers has
stressed:

> 'Poor people on disaster courses may not be recognized. A nutritionist may see
> malnutrition but not the seasonal indebtedness, the high cost of medical treat-
> ment, the distress sales of land, and the local power structure which generate it.
> A doctor may see infant mortality but not the declining real wages which drive
> mothers to desperation, still less the causes of those declining real wages.
> Visibility and specialisation combine to show simple surface symptoms rather
> than deeper combinations of causes. The poor are little seen, and, even less is
> the nature of their poverty understood.'(Chambers 1983, p. 25.)

Many of the case studies presented have recognized this need to understand the nature of a community's poverty and the social and political causes
of their health problems. This has often led to programmes broadening their
approach to political lobbying and income-generating activities, and a re-
appraisal of what needs to be done to improve health. With the need to act
upon the many different underlying causes of ill health, there is thus a need
for 'multi-sectoral' action.

Multi-sectoral action

There is often a multitude of agencies and institutions on the scene in the

Fig. 15.3   The children of poor urban communities are most at risk. Photograph by C. Goyder.

urban environment. It is particularly true in urban environments that the linkages among social, environmental and economic conditions, individual and community behaviour, health status and general well-being are so intricate that there has to be some attempt at intersectoral co-ordination and collaboration. Although difficult, the municipality can sometimes fulfill this role (see for example the Hyderabad case study in Ch. 8).

Where co-ordination is poor the effectiveness of different projects on health may be counter-productive. This difficulty has been mentioned in a number of case studies. Community health workers (CHWs) can also fulfill an intersectoral role by taking on community development work, which can involve work in, for example, housing, education, water and sanitation, and nutrition. There is a need for more pragmatically trained CHWs who are integrated in community-oriented multi-disciplinary health teams and linked to strong support and referral networks. Although urban CHWs may be losing out on some cash income, and it is therefore harder to recruit them and to pay them in kind than their rural counterparts, we have had many examples of successful urban CHWs (Ch. 8 and 13).

Co-ordinating mechanisms involving health, education, public works, industry, commerce, housing, transport, communications, and other relevant departments are needed. Probably the most successful case studies that we know of are those which have taken a multi-sectoral perspective and addressed problems in many of these spheres. However, most of these fully-integrated programmes were run by NGOs. Examples at a government level have usually come across the kind of problems experienced by the attempts at intersectoral collaboration in Kuala Lumpur (Ch. 12), although the Hyderabad Urban Community Development programme seems to be an exception.

In addition to intersectoral co-ordination there is a need for more intrasectoral co-ordination within the health sector in the city. Often the municipality and the ministry of health have no mechanisms to co-ordinate their actions. Good examples of co-ordination within the health sector, involving hospitals, clinics, and CHWs are rare.

The role of the hospitals

One of the most obvious differences between rural and urban health services is that urban areas have hospitals which the rural areas lack. There is an increasing movement to involve hospitals in primary health care. A raising of awareness, within the services and outside, should, in turn, lead to improved and more community-oriented hospital services, medical schools, and nursing schools. In the city one may find municipal hospitals, ministry of health hospitals, private, mission, and social security hospitals. With the strength of manpower behind such institutions the potential contribution to urban primary health care is great. We need to learn more about how hospitals can link with communities, using examples such as the Aga Khan University in Karachi, Pakistan, which has linked hospitals and communities by placing medical doctors in selected slums for training periods and then providing follow-up services.

Collaboration between government and non-governmental agencies

In addition to different sectors or disciplines needing to collaborate, there is a need for more liaison between non-governmental and governmental organizations. There are few good examples of such collaboration. Often what is needed is an intermediary or 'broker' of some sort. UNICEF, for example, fulfilled this role in urban primary health care in the Philippines (Rialp 1985). Although UNICEF in the Philippines works almost exclusively with the government and supports government programmes, it encouraged the government to include NGO projects under the Urban Basic Services Programme. The government's decision to finally approve the NGO projects for UNICEF support was coloured by the recognition that the innovative approaches of NGOs to urban community development could provide important lessons to be learned from which could be drawn into a more broadly based (i.e. government, national) strategy.

When collaboration like this succeeds the government learns from the NGO experience and the NGOs link into the government's service delivery network. Therefore, scaling up urban primary health care becomes more efficient and cost-effective (see Ch. 14). Often it is the NGOs who must take the initiatives. The involvement of an intermediary can often legitimize groups and programmes ordinarily unacceptable to governments. On the other hand, an intermediary can often break down the stereotypical anti-government stance that some NGOs have by arranging contacts and shared activities.

Information needs

The lack of data on the health conditions of the urban poor has been emphasized throughout this book. Health programmes in poor urban communities often fail to produce any good data about their members, and no evaluation of the programme is carried out. The most important reason for collecting health data should not be for the benefit of national and international bodies who want to be informed of the progress of a country towards Health for All by the Year 2000, but to inform the communities and their health personnel. A lack of local information often leaves people in the dark about the problems confronting their programmes and how they might be resolved.

Several of the case studies presented here have implemented baseline surveys in order to chart the progress of a programme. Community health workers have been used to monitor components for evaluation, through the use of family health cards. This can be done by government and NGO programmes alike. For example, the Manila City Health Department was covering 2095 low-income families in 1984 with a lay-reporting mechanism, with the aim of developing a health information monitoring system for the urban primary health care programme (Suva 1985). However, very few in-depth evaluation studies have been done to document the impact of these urban programmes on health and social status.

Turning to more 'academic' research, Chapter 4 discussed the need for more studies that examine intra-urban differentials, i.e. breaking down the misleading city averages that are currently available. In addition to data which point to the differences in health between middle-income and low-income sectors of the city, more research is needed to shed light upon the difference between low-income communities and within low-income communities. For example, do recent migrants have different health problems compared to more stable residents? High-risk groups need to be identified so that health action in favour of these groups can be planned and monitored.

Fig. 15.4   Street children remain one of the major challenges when it comes to urban health care. Photograph by OXFAM.

Health action itself can also be arranged in order of priority according to major health problems.

In addition to studies which focus upon health status, there is a need for research into the various processes involved in implementing urban primary health care. Case studies, including those in this book, can provide single examples of processes but there is a need for special comparative studies on such topics as:

- how multi-sectoral policies and plans relevant to PHC can be implemented in the city,
- the mechanisms for effective intersectoral co-ordination and collaboration,
- the management of PHC in cities,
- the utilization of health services in the city,
- appropriate community involvement,

- income generation and its role in alleviating urban poverty,
- the role of child care facilities in the promotion of children's health,
- abandoned children and their particular health problems,
- the study of urban disabled children,
- accessibility to water and sanitation,
- the relationship between housing and health in the poor urban environment,
- the influence of traditional medicine on the health of the population,
- financing and costs of urban PHC.

The results of such evaluations and research studies can also be very useful in training programmes. Yesudian *et al.* (1985) have described how such programmes can be particularly effective where people from different backgrounds, e.g. hospital managers, city council officials, politicians, community representatives, academics, and community workers meet to devise suitable strategies for urban primary health care.

The need for an urban health policy

Several policy implications can now be identified. The first concerns reorienting urban health services. Urban primary health care cannot succeed without some fundamental changes in attitude and approach within health systems and within government, since it cannot simply be an 'add on' to existing services. Primary health care has a special relevance and urgency in the poorest urban neighbourhoods, where health is much worse than most people recognize, even though the main resources of the health care system are nearby. Many governments have recognized the potential of primary health care in rural areas. With rapidly growing numbers of poor urban people it is imperative that governments review the structures, facilities, and methodologies available within their country for urban primary health care. Strong advocacy is needed to explain and gain acceptance for urban primary health care.

There are few examples of a truly comprehensive city health plan or of a systematic managerial process tackling the problems of urban deprivation. Nevertheless, cities such as Addis Ababa, Manila, Mexico City, and the many cities involved in the Urban Community Development programmes in India are moving in the right direction and reach out in their planning to vulnerable groups. However, the present economic recession in certain cities, such as Mexico City, jeopardizes the large-scale implementation of these plans. The bulk of the money to pay for urban primary health care must come through re-allocation of resources in the health system and elsewhere in public budgets. None of the case studies presented here were totally financed by the community itself. Communities may pay a token amount or give their time but external resources are always needed. Such resources may come from local NGOs, the municipality, government sectors, international

NGOs, multi-lateral agencies, etc. The available evidence suggests that US$1–5 per head per annum will pay for comprehensive primary health care, in both rural and urban areas. One suggestion is that in countries where money is allocated to cities on a programme basis, central and provincial governments could make statutory provision to insist that cities spend at least a defined proportion of their budgets on PHC programmes (WHO 1986).

Legislation can sometimes inhibit the improvement of the health and living conditions of the urban poor, for example, by prescribing unrealistically high standards for housing or unrealistically low housing densities. Often these standards were, initially, established for *élites* in developed countries according to criteria that are not appropriate to the slums of cities of developing countries (England 1981). On the other hand, it is sometimes lack of legislation that is the problem. One of the most useful actions that authorities can take is the granting (or regularizing) of land

Fig. 15.5 It is only when communities organize themselves that the residents can look out towards a brighter future. Photograph by A. Charnock.

tenure in illegal settlements. Once an illegal settlement is legalized, there is a stronger case for the public provision of basic infrastructure and services. There is a need to emphasize slum and squatter area upgrading rather than the traditional policies of eviction. Such upgrading schemes (like those implemented by the World Bank) could form useful starting points for health-related activities. This kind of planning is also likely to reflect the felt needs of the population, in that physical improvements are often a priority.

When it comes to deciding which population groups should receive priority, the argument is similar to that put forward for equity for rural populations. If additional resources become available, they should be allocated preferentially to meet the needs of deprived groups. If no additional resources can be obtained, a re-allocation of what is already available within the urban health system will have to be undertaken, in order to achieve a more equitable and more effective distribution of services (Rossi-Espagnet 1985).

Primary health care among the urban poor has great potential for winning political support from the people. Through a process of involving and organizing them to overcome health problems, the groundwork is also laid for other socio-economic and political development activities (Shubert 1986). We are gradually learning more about the processes and problems associated with managing and implementing urban primary health care. It is now up to municipalities, governments, non-governmental organizations, and international agencies to support the urban poor in their attempt to emerge from the Shadow Of The City.

References

Adamson, P. (1982). The Gardens. *New Internationalist* **109**, 7–28.

Aga Khan Foundation (1981). *The role of hospitals in Primary Health Care*. Report of a conference sponsored by the Aga Khan Foundation and the World Health Organization, 22–26 November 1981, Karachi, Pakistan.

Agnelli, S. (1986). *Street children: a growing urban tragedy*. A report for the Independent Commission on International Humanitarian Issues. Weidenfeld Paperbacks.

Amis, P. (1982). *Squatters or tenants: the commercialization of unauthorized housing in Nairobi*. Paper presented at the Development Studies Annual Conference in Dublin, September 1982.

Arias, J. (1977). Health care in big cities, Bogota. *World Hospitals* **13**, 83–90.

Assignment Children (1983). Some facts and figures on ORT. Editorial. 61/62,69.

Austin, J.E. (1980). *Confronting urban malnutrition: the design of nutrition programmes*. World Bank Staff Occasional Papers, No. 28. John Hopkins University Press, Baltimore.

Bapat, M. and Crook, N. (1984). The environment, health and nutrition: an analysis of inter relations from a case study of hutment settlements in the city of Poona. *Habitat International* **8**, 115–26.

Basta, S.S. (1977). Nutrition and health in low income urban areas of the Third World. *Ecology of Food and Nutrition* **6**, 113–24.

Bennagen, P.L. (1981). *Primary Health Care in the city of Manila: a case study*. Report of the WHO Regional Seminar on Urban Primary Health Care, Manila, 30 November–4 December 1981.

Bennett, F.J. (1979). *Community diagnosis and health action*. A manual for tropical and rural areas. Macmillan, London.

Benyoussef, A., Cutler, V.L., Baylet R., Collomb, H., and Drop, S. (1973). Santé, migration et urbanization, une étude collective au Senegal. *Bulletin of the World Health Organization* **49**, 517–37.

Berg, A. (1981). *Malnourished people: a policy view. Poverty and basic needs series*. The World Bank, Washington DC.

Bianco, M. (1983). *Health and it's care in the greater Buenos Aires*. Paper prepared for the joint UNICEF/WHO meeting on Primary Health Care in Urban Areas, 25–29 July 1983. WHO document SHS/HSR/83.1, Geneva.

Boonyabancha, S.(1983). The causes and effects of slum eviction in Bangkok. In *Land for housing the poor* (ed. Angel, Archer, Tanphiphat, and Wegelin). Select Books, Singapore.

Bosnjak, V. (1985). *Scaling up of local poverty programmes*. A paper prepared for a workshop on Community health and the urban poor, organized by the London

School of Hygiene and Tropical Medicine, OXFAM, and UNICEF, 7–12 July 1985, Oxford, England.

Bosnjak, V. and Cayon, E. (1985). *Implementation of Primary Health Care services with community participation in the city of Coatzacoalcos State of Veracruz, Mexico*. Paper prepared for a workshop on Community health and the urban poor, organized by the London School of Hygiene and Tropical Medicine, OXFAM, and UNICEF, 7–12 July 1985, Oxford, England.

Boulos, R. (1985). *Cité Simone, an urban slum in Haiti: the malnutrition challenge.* Paper prepared for a workshop on Community health and the urban poor, organized by the London School of Hygiene and Tropical Medicine, OXFAM, and UNICEF, 7–12 July 1985, Oxford, England.

Brandberg, B. (1983). *The latrine project, Mozambique.* IDRC Manuscript Reports no. IDRC-MR58e. International Development Research Centre, Ottawa.

Brink, E.W., *et al.* (1983). The Egyptian national nutrition survey 1978. *Bulletin of the World Health Organization* **61**, 853–60.

Brown, R. (1985). *Operation Friendship.* A paper presented for a workshop on Community Health and the urban poor, organized by the London School of Hygiene and Tropical Medicine, OXFAM, and UNICEF, 7–12 July 1985, Oxford, England.

Brownrigg, G. (1985). *Home gardening in international development: what the literature shows.* League for International Food Education, Washington DC.

Bunnag, J.E. (1981). Communicating for health: a third world perspective. In *Health education and the media.* (ed. Leathar, D.S. *et al.*). Pergamon, Oxford.

Burton, J.H. (1976). Problems of child health in a Peruvian shanty town. *Tropical Doctor* **6**, 81–3.

Cairncross, S. (1985). *Sanitation and the urban poor.* Paper prepared for a workshop on Community health and the urban poor, organized by the London School of Hygiene and Tropical Medicine, OXFAM, and UNICEF, 7–12 July 1985, Oxford, England.

Cairncross, S. and Feachem, R. (1983). *Environmental health engineering in the tropics: an introductory text.* Wiley, Chichester.

Carriconde, C. (1985). *Casa Amarela community health project, Recife, Brazil.* Paper prepared for a workshop on Community health and the urban poor, organized by the London School of Hygiene and Tropical Medicine, OXFAM, and UNICEF, 7–12 July 1985, Oxford, England.

Carroll, A. (1980). *Pirate sub-divisions and the market for residential lots in Bogota.* World Bank staff working paper no. 435. World Bank, Washington DC.

Cassim, J.K., Peries, T.H.R., Jayasinghe, V., and Fonseka, L. (1982). Development councils for participatory urban planning—Colombo, Sri Lanka. *Assignment Children* **57/58**, 157–90.

Centre for Science and Environment, New Delhi (1985). The State of India's environment 1984–85: *The second citizen's report.*

Chambers, R. (1983). *Rural development—putting the last first.* Longman, London.

Chen, M. (1983). *The working women's forum: organizing for credit and change.* SEEDS publication No. 6. Population Council, USA.

Chernichovsky, D. and Meesook, O.A. (1985). *Urban–rural food and nutrition* consumption patterns in Indonesia. Population, Health and Nutrition Technical Note 85-5. Population, Health and Nutrition Department. World Bank, Washington DC.

Choquehuanca, V. (1985). *Emergency health and nutrition programme of the Metropolitan Municipality of Lima, Peru.* Paper prepared for a workshop on Community health and the urban poor, organized by the London School of Hygiene and Tropical Medicine, OXFAM, and UNICEF, 7–12 July 1985, Oxford, England.

Chowdhury, Z. (1981). The good health worker will inevitably become a political figure. *World Health Forum* **2**, 55–7.

Chung, K.K. (1980). *Pattern of utilization of health care by the urban poor.* Korean Health Development Institute, Seoul, Korea.

Connolly, P. (1982). Uncontrolled settlements and selfbuild: What kind of solution? The Mexico City case. In *Self-help housing: a critique* (ed. P. Ward). Mansell, UK.

Corry, M. (1985). *Development of a Community Health Programme—The Cape Town Centre of St. John's Ambulance.* Paper prepared for a workshop on Community health and the urban poor, organized by the London School of Hygiene and Tropical Medicine, OXFAM, and UNICEF, 7–12 July 1985, Oxford, England.

Coulibaly, N. (1981). Place et approches des problémes de la tuberculose en Abijian. *Medicine Afrique Noire* **28**, 447–9.

Cousins, W.J. and Goyder, C. (1986). Hyderabad Urban Community Development Project. In *Reaching the urban poor: project implementation in developing countries* (ed. G. Shabbir Cheema), pp. 181–206. Westview Press, Boulder.

Creed, H. (1985). *Urban nutrition: a case study. The health and nutrition programme, Pamplona Alta, Lima, Peru.* Paper prepared for a workshop on Community health and the urban poor, organized by the London School of Hygiene and Tropical Medicine, OXFAM, and UNICEF, 7–12 July 1985, Oxford, England.

Cuenya, B., Almada, H., Armus, D., Castells, J., di Loreto, M., and Penalva, S. (1984). *Habitat and health conditions of the popular sector; a pilot project of participative investigation in the San Martin settlement, Greater Buenos Aires.* Centro de Estudios Urbanos y Regionales, Buenos Aires.

Datta Banik, N.D. (1977). Some observations on feeding programmes, nutrition and growth of pre-school children in an urban community. *Indian Journal of Pediatrics* **44**, 139–49.

Datta Banik, N.D. (1978). Epidemiology of gastroenteritis of pre- school children in slum areas in Delhi with reference to helminthic and parasitic infection. *Indian Journal of Pediatrics* **45**, 303–9.

Davidson, J. (1983). The survival of traditional medicine in a Peruvian Barriada. *Social Science and Medicine* **17**, 1271–80.

Davis, I. (1984). The squatters who live next door to disaster. *Guardian Third World Review,* 7 December 1984.

de Carvalho, J.A. and Wood, C.H. (1978). Mortality, income distribution and rural–urban residence in Brazil. *Population and Development Review* **4**, (3) 405–20.

de la Paz, T. (1985). *The selection, training, role and support system of the Katiwala— a volunteer health worker.* Paper prepared for a workshop on Community health and the urban poor, organized by the London School of Hygiene and Tropical Medicine, OXFAM, and UNICEF, 7–12 July 1985, Oxford, England.

Desai, A.R. and Pillai, S.D. (1972). *A profile of an Indian slum* Bombay University Press, Bombay.

Dias, L., Camarano, M.R., and Lechtig, A. (1985). *Drought, recession and prevalence of low birthweight babies in poor urban populations of the northeast of Brazil*. UNICEF, Brazil.

Diskett, P. (1986). *Brazil Nutrition Survey*. OXFAM UK. Mimeograph.

Donohue, J.J. (1982). Facts and figures on urbanization in the developing world. *Assignment Children* **57/58**, 21–41.

Ebrahim, G.J. (1983). Primary health care and the urban poor. Editorial. *Journal of Tropical Paediatrics* **29**, 2–3.

Ebrahim, G.J. (1984). Health care and the urban poor. In *Basic needs and the urban poor: the provision of communal services*. (ed. P.J. Richards *et al.*) Croom Helm, London.

Egdell, H.G. (1983). Mental health care in the developing world. *Tropical Doctor* **13**, 149–52.

Eisenberg, C. (1980). Honduras: mental health awareness changes a community. *World Health Forum* **1**, (1 + 2), 72–7.

Engels, F. (1845). *The condition of the working class in England*. Translated from the German edition. Panther 1974.

England, R. (1981). Legislation prevents progress. *Consulting Engineer*, 20–21 October 1981.

Escola Paulista, (1975). *Estado Nutricional de Criancas de 6 a 60 meses no Municipio de Sao Paulo*, Vol. 2. Ministerio de Educacao e Cultura, Sao Paulo.

Fajardo, O.G. (1983). *The health services of Mexico City for people living in marginal areas*. Paper presented for the UNICEF/WHO joint meeting on Primary Health Care in Urban Areas, 25–29 July 1983. WHO document SHS/HSR/83.1, Geneva.

Family Planning Foundation, India (1984). *Working women's forum: experiment in leadership training that blazes a trail*. Paper presented at the International Conference on Population, Mexico City.

Finquelievich, S. (1985). *Alternative policies to improve access to food, energy and shelter for low-income families in Latin America Metropolis*. Paper presented at an international conference on Families in the face of urbanization, organized by IUFO and UNICEF, 2–5 December 1985, New Delhi, India.

Flintoff, F. (1976). *Management of solid wastes in developing countries*. World Health Organization South East Asia Region, New Delhi.

Fonseka, L. (1985). *Urban basic services in India: an attempt at operational expansion*. Paper presented at UNICEF Urban Basic Services Workshop, 13–17 October, Karachi.

Frank, A.G. (1971). *Capitalism and underdevelopment in Latin America*. Penguin, Harmondsworth.

Ganapati, R. (1983). Urban leprosy control. *Tropical Doctor* **13**, 76–8.

Ganapati, R., Naik, S.S., Acharekar, M.Y., and Pade, S.S. (1976). Leprosy endeminity in Bombay: an assessment through surveys of municipal schools. *Leprosy Review* **47**, 127–31.

Gelfand, M. *et al.* (1981). The urban N'Arga in practice. *Central Africa Journal of Medicine* **27**, 93–5.

Getulio Vargas Foundation (1975). *Pesquisa sobre consumo alimentar.* Brazilian Institute of Economics, Rio de Janeiro.

Good, C.M. and Kimani, V.N. (1980). Urban traditional medicine: A Nairobi case-study. *East African Medical Journal* **57**, 301–16.

Goyder, C. (1985). *Urban health care in Addis Ababa—a case study.* Paper prepared for a workshop on Community health and the urban poor, organized by the London School of Hygiene and Tropical Medicine, OXFAM, and UNICEF, 7–12 July 1985, Oxford, England.

Gracey, M., Stone, D.E., Sutoto, S., and Sutejo, S. (1976). Environmental pollution and diarrhoeal disease in Jakarta, Indonesia. *Journal of Tropical Paediatrics* **22**, 18–23.

Griffith, M.H. (1983). Urban health workers in search of a role. *Future* **1**, 43–7.

Guimaraes, J.J. and Fischmann, A. (1985). Inequalities in 1980 infant mortality among shantytown residents and non-shantytown residents in the municipality of Porto Alegre, Rio Grande do Sul, Brazil. *Bulletin of the Pan American Health Organization* **19**, 235–51.

Hailu, Y. (1978). *Upgrading the Teklehaimonot: existing conditions and proposed development.* World Bank, Washington DC.

Hamer, A.M. (1981). *Bogota's unregulated subdivisions: The myths and realities of incremental housing construction.* Urban and Regional Report No. 81–19. World Bank, Washington, DC.

Hardie, M. (1985). *City hospital support for community health care.* Paper prepared for a workshop on Community health care and the urban poor, organized by the London School of Hygiene and Tropical Medicine, OXFAM, and UNICEF, 7–12 July 1985, Oxford, England.

Hardoy, J.E. and Satterthwaite, D.E. (1984). *Third World cities and the environment of poverty.* Background paper for the World Resources Institute 'Global Possible' Conference, 9 March 1984.

Hardoy, J.E. and Satterthwaite, D.E. (1985*a*). Third World Cities and the environment of poverty. In *The global possible: resources development and the new century.* (ed. R. Repetto) pp. 171–210. Yale University, USA.

Hardoy, J.E. and Satterthwaite, D.E. (1985*b*). *Shelter, infrastructure and services in Third World cities.* Paper prepared for the Forum Meeting of the World Commission on Evironment and Development, Sai Paulo, Brazil, 25 October–4 November.

Hardoy, J.E. and Satterthwaite, D.E. (1986*a*). *Small and intermediate urban centres in the Third World; their role in national and regional development.* Hodder and Stoughton, London.

Hardoy, J.E. and Satterthwaite, D.E. (1986*b*). Urban change in the third world: Are recent trends a useful pointer to the urban future? *Habitat International* **10**(3), 33–52.

Harpham, T. (1986). Review article: health and the urban poor. *Health Policy and Planning* **1**, 5–18.

Hasan, A. (1985). *The low cost sanitation programme of the Orongi Pilot Project, Karachi Pakistan.* Paper prepared for a workshop on Community health and the urban poor, organized by the London School of Hygiene and Tropical Medicine, OXFAM, and UNICEF, 7–12 July 1985, Oxford, England.

Heggenhougen, H.K. (1980). Bohoms, doctors and sinsehs—medical pluralism in Malaysia. *Social Science and Medicine* **14B**, 235–44.

Heggenhougen, H.K. (1984). Will Primary Health Care efforts be allowed to succeed? *Social Science Medicine* **19**, 217–24.

Heggenhougen, H.K., Vaughan, J.P., Muhondwa, E., and Rutabanzibwa-Ngaiza, J. (1987). *Community Health Workers: the Tanzanian experience.* Oxford University Press, London.

Herbert, J. and Teas, J. (1985). Urban third world children: toxic exposure and malnutrition. *Science for the People* November/December, 18–22.

Hinkle, L.E. and Loring, W.C. (1977). *The effect of the man-made environment on health and behaviour.* Centre for Disease Control, Public Health Service, Atlanta.

Hirst, N. (1983). Bricoleurs, Brohers, mediators, healers and magicians: the diviners role and function in Grahamstown (South Africa). *Curare* **6**, 51–68.

Hobcraft, J. (1985). World Fertility Survey: a final assessment. Survey throws new light on key policy issues. *People* **12**, 3–5.

Holmes, J.R. (ed.) (1984). *Managing solid wastes in developing countries.* Wiley, Chichester.

Imperato, P.V. (1979). Traditional medical practitioners among the Bambara of Mali and their role in the modern health care delivery system. In *African Therapeutic Systems*, ed. Ademuwagun, Aycade, Hamson, and Warren. Crossroads Press USA.

Ishikawa, N. and Nabi, G. (1981). *Estimation of the current risk of tuberculosis infection in Bangladesh.* Proceedings of the 12th Eastern Regional Tuberculosis Conference of the International Union Against Tuberculosis, 7–12 December 1981, Dhaka, Bangladesh.

Jere, H. (1985). *Experiences in urban community mobilization and involvement in development projects in Lusaka.* Paper prepared for a workshop on Community health and the urban poor, organized by the London School of Hygiene and Tropical Medicine, OXFAM, and UNICEF, 7–12 July 1985, Oxford, England.

Jha, S.S. (1985). *Urban health in underdeveloped countries with special reference to women and children.* Paper presented at an international conference on Families in the face of urbanization, sponsored by IUFO and UNICEF, 2–5 December 1985, New Delhi, India.

Johnson, G. (1964). Health conditions in rural and urban areas of developing countries. *Population Studies* **17**, 293–309.

Kalibermatten, J.M., Julius, D.S., and Gunnerson, C.G. (1980). *A sanitation field manual.* Appropriate technology for water and sanitation No. 11. The World Bank, Washington DC.

Karamoy, A. (1984). The kampung improvement program: Hope and reality. *Prisma, the Indonesian Indicator* **32**, 19–36.

Kelley, A.C. and Williamson, J.G. (1984). Population growth, industrial revolutions and the urban transition. *Population and Development Review* **10**, 419–41.

Kerejan, H. and N'Da Kowan (1981). Approches des problémes alimentaires et nutritionnes d'une megalpolis africaine. *Medicine Afrique Noire* **28**, 479–82.

Keyes, W.J. (1980). Metro Manila, The Philippines. In *Policy towards urban slums: slums and squatter settlements in the ESCAP region—case study of seven cities* (ed. Sarin M.). ESCAP United Nations, New York.

Khanjanasthiti, P. and Wray, J.D. (1974). Early protein–calorie malnutrition in slum areas of Bangkok Municipality 1970–1971. *Journal of Medical Association Thailand* **57**, 357–66.

Kingma, S. (1977). Finding the community in our crowded cities. *Contact* **38**, 1–2.

Kleevens, J.W.L. (1966). Rehousing and infections by soil transmitted helmintins in Singapore. *Singapore Medical Journal* **7**, 12–29.

Korten, D. (1980). Community organization and rural development: A learning process approach. *Public Administration Review* September/October, 480–511.

Kothari, G. Gandenver, K., and Rumbeck J. (1983). *Appraisal of current conditions of health in greater Bombay*. Paper presented at the seminar on Problems of public health in metropolitan cities, International Institute of Population Studies, Bombay.

Kouray, M and Vasquez, M.A. (1979). Housing and certain socio-environmental factors and prevalence enteropathogenic bacteria among infants with diarrhoeal disease in Panama. *American Institute of Tropical Medicine and Hygiene* **18**, 936–41.

Lee, One-fu and Furst, B.B. (1980). *Differential indicators of living conditions in urban and rural places of selected countries*. Applied Systems Institute, Washington DC.

Lewis, O. (1965). *La vida: a Puerto Rican family in the culture of poverty*. Secher and Warburg, London.

Lewis, W.J., Foster, S.S.D., and Drasar, B.S. (1980). *The risk of groundwater pollution by on-site sanitation in developing countries: a literature review*. IRCWD Report No. 01/82. International Reference Centre for Wastes Disposal, Duebendorf.

Lipton, M. (1976). *Why poor people stay poor: urban bias in world development*. Harvard University Press, Cambridge, Mass.

Lopez, B., Brown, K.H., and Black, R.E. (1985). Survey of health conditions and nutritional status of infants and young children in Huascar, an underprivileged peri-urban community of Lima, Peru. Unpublished document.

Lusk, G. (1982). Sudan's three towns: capital struggles with unwieldy sanitation programme. *World Water* **5**(11), 47–57.

Macagba, R.L. (1984). *Hospitals and Primary Health Care*. International Hospitals Federation, London.

M'Gonigle, G.C.M. (1933). Poverty, nutrition and the public health. An investigation into some of the results of moving a slum population to the modern dwellings. *Proceedings of the Royal Society Medicine* **26**, 677–87.

Mahaniah, J. (1981). The multidimensional structure of healing in Kinshasa, capital of Zaire. *Social Science Medicine* **15B**, 341–49.

Mahler, H. (1982). *The new look in health education.* Paper for Expert committee on new approaches to health education in Primary Health Care. WHO, Geneva.

Marga Institute (1982). *Colombo City information: a Marga Institute compilation* Doc. No. M/692.

Mason, J.P. and Stephens, B. (1981). *Housing and health: An analysis for use in the planning, design and evaluation of low-income housing problems.* Office of Housing, USAID, Washington DC.

Maxwell, R.J. (1986). Health care in urban areas. *The Lancet* **1**, 222–3.

Mayer, K. (1985). *Community Health Care Programme in low-income areas in Santiago, Chile.* Paper prepared for a workshop on Community health and the urban poor, organized by the London School of Hygiene and Tropical Medicine, OXFAM, and UNICEF, 7–12 July 1985, Oxford, England.

Mohan, R., Garcia, J., and Wagner, M.W. (1981). *Measuring urban malnutrition and poverty: a case study of Bogota and Cali, Colombia.* World Bank Staff Working Paper No. 447. World Bank, Washington DC.

Monckeberg, F. (1968). Efecto de la nutricion medio zubiente sobre el desarolo psico—motor en el nijo. *Caudarnos Medico—Sociales* **9**(5).

Morell, S. and Morell, D. (1972). *Six slums in Bangkok: problems of life and options for actions.* UNICEF, Bangkok.

Moser, C.O.N. (1977). The dual economy and marginality debate and the contribution of micro-analysis; market sellers in Bogota. *Development and Change* **8**, 465–89.

Moser, C.O.N. (1978). Informal sector or petty commodity production: dualism or dependence in urban development. *World Development* **6**, 9–10.

Moser, C.O.N. (1980). Why the poor remain poor: the experience of Bogota Market Traders in the 1970s. *Journal of Interamerican Studies and World Affairs* **22**, 3.

Moser, C.O.N. (1982). A home of one's own: squatter housing strategies in Guayaquil, Ecuador. In *Urbanization in Contemporary Latin America* (ed. Gilbert, A. *et al.*) pp. 159–90. Wiley, London.

Moser, C. and Satterthwaite, D.E. (1985). *The characteristics and sociology of poor urban communities.* Paper prepared for a workshop on Community health and the urban poor, organized by the London School of Hygiene and Tropical Medicine, OXFAM, and UNICEF, 7–12 July 1985, Oxford, England.

Muller, F. (1983). Contrasts in community participation. In *Practising health for all*, Morley, D., Rohde, J. and Williams G. (ed. Morley *et al.*). Oxford University Press, Oxford.

Munro, I. (1986). *Primary health care and shelter policies in developing countries.* Paper presented at the UNICEF/WHO Inter-regional Consultation on Primary Health Care in Urban Areas, Manila, Philippines, 7–11 July 1986. Report No. SHS/IHS/86.1, WHO Geneva.

Musinde, S. and Lamboray, J.L. (1985). *Primary Health Care in an urban area in Kinshasa, Zaire: health for all project Kinshasa* Paper prepared for a workshop on Community health and the urban poor, organized by the London School of

Hygiene and Tropical Medicine, OXFAM, and UNICEF, 7–12 July 1985, Oxford, England.

Myers, R.G. (1984). *Going to scale*. A paper prepared for the Second Inter-agency Meeting on Community-based child development, 29–31 October, New York.

Nelson, V. and Mandl, P.E. (1978). Peri-urban Malnutrition. A neglected problem. *Assignment Children* **43**, 25–46.

Newell, K. (1985). *Urban community health workers*. Paper prepared for a workshop on Community health and the urban poor, organized by the London School of Hygiene and Tropical Medicine, OXFAM, and UNICEF, 7–12 July 1985, Oxford, England.

Nitaya, C. and Ocharoen, U. (1980). Bangkok, Thailand. In *Policies toward urban slums: slums and squatter settlements in the ESCAP region—case studies of seven cities* (ed. Sarin M.). ESCAP United Nations, New York.

Njau, G. (1982). Human settlements in the 1980's: Nairobi's experience. In *Papers and proceedings of Habitat Forum Conference* 1980/81 (ed. Blair).

Nyi Nyi. (1984). Going to scale: Going national—operationalization process and issues. *Assignment Children* **65/68**, 23–36.

Odejide, A.O. Olatawura, M.O. *et al.* (1977). Traditional healers and mental illness in the City of Ibadan. *African Journal of Psychiatry* 99–106.

Oluwande, P.A., Sridhar, M.K.C. and Okubadeuo, O. (1978). The health hazards of open drains in developing countries. *Progress in Water Technology* **11**, 121–30.

Ongari, L.A. and Schroeder, F. (1985). *Community based health care programme of the Undugu society of Kenya*. Paper prepared for a workshop on Community Health and the urban poor, organized by the London School of Hygiene and Tropical Medicine, OXFAM, and UNICEF, 7–12 July 1985, Oxford, England.

OXFAM (1986). Oxfam file number SDN203, Ockenden Venture, Refugee Training in Port Sudan. OXFAM, Oxford, England.

Pacey, A. (1978). *Sanitation in developing countries*. Wiley, Chichester.

Parikh, I. (1985). *Streehitakarini urban community health programme—part of an integrated development programme*. Paper prepared for a workshop on Community health and the urban poor, organized by the London School of Hygiene and Tropical Medicine, OXFAM, and UNICEF, 7–12 July 1985, Oxford, England.

Pascual, C.J. (1984). *Urban community based rehabilitation services*. Paper presented at the joint UNICEF/WHO consultation on Primary Health Care in urban areas, 15–19 October 1984. WHO report SHS/84.5, Guayaquil.

Patel, I. 1985. *Design for human dignity—a case study for the problem of Indian latrines*. Paper prepared for a workshop on Community health and the urban poor, organized by the London School of Hygiene and Tropical Medicine, OXFAM, and UNICEF, 7–12 July 1985, Oxford, England.

Paul, S. (1982). *Managing Development Programs: the lessons of success*. Westview Press, Boulder, Colorado.

Peattie, L. (1975). Tertiarization and urban poverty in Latin America. *Latin American Urban Research* **5**, 109–23.

Perera, L.N. (1972). *The effects of income on food habits in Ceylon: the findings of the socio-economic survey.* Cornell agricultural economic staff paper, Ithaca.

Peries, T. (1985). *Community health and the urban poor, Colombo, Sri Lanka.* Paper prepared for a workshop on Community health and the urban poor, organized by the London School of Hygiene and Tropical Medicine, OXFAM, and UNICEF, 7–12 July 1985, Oxford, England.

Perlman, J. (1976). *The myths of marginality: urban poverty and politics in Rio de Janerio.* University of California Press, Berkley.

Pinto, G. (1985). *Primary Health Care in urban areas with reference to programmes by the Bombay Municipal Corporation and voluntary agencies.* Paper prepared for a workshop on Community health and the urban poor, organized by the London School of Hygiene and Tropical Medicine, OXFAM, and UNICEF, 7–12 July 1985, Oxford, England.

Plank, S.V. and Milanesi, M.L. (1973). Infant feeding and mortality in rural Chile. *Bulletin of World Health Organization* **48**, 203.

Poerbo, H., Sicular, D.T., and Supardi, V. (1984). An approach to development of the informal sector: the case of garbage collectors in Bandung. *Prisma, the Indonesian indicator* **32**, 85–101.

Portes, A. (1978). *The informal sector and the world economy: notes on the structure of subsidized labour.* Institute of Development Studies Bulletin No. 9. Sussex University, Brighton.

Potts, M. and Bhiwandiwala, P.P. (1981). *Meeting the family planning needs of the urban poor.* International Fertility Research Programme, Research Triangle Park, North Carolina.

Prasada Rao, T.M., Gowrinath-Sastry, J., and Vijayarghansan, K. (1974). Nutritional status of children in urban slums around Hyderabad City. *Indian Journal of Medical Research* **62**, 1492–8.

Puffer, R.R. and Serrano, C.V. (1973). *Patterns of mortality in childhood.* Pan American Health Organization Science Publication 262, Washington DC.

Pyle, D. (1984). *Life after project (a multi-dimensional analysis of implementing social development programs at the community level).* John Snow, Boston.

Ramasubban, R. and Crook, N. (1985). The mortality toll of cities: emerging patterns of disease in Bombay. *Economic and Political Weekly (Bombay)* **20**(23), June 8.

Rau, R. (1985). *Urban Community Development Project—Hyderabad.* Paper prepared for a workshop on Community health and the urban poor, organized by the London School of Hygiene and Tropical Medicine, OXFAM, and UNICEF, 7–12 July 1985, Oxford, England.

Rego, R. and Cuentro, S. (1984). *Experiencia de saneamento de baixo custo em Olinda.* Seminar on alternative technologies in low-cost sanitation, Recife, Brazil.

Rialp. V. (1985). *Enhancing NGO-government collaboration in urban primary health care: three Phillipine NGO experiences.* Paper prepared for a workshop on Community Health and the Urban Poor, organized by the London School of Hygiene and Tropical Medicine, OXFAM, and UNICEF, 7–12 July 1985, Oxford, England.

Richardson, N.J.H., Hayden-Smith, S., Bokkenheuser, V., and Koornhof, H.J. (1968). Salmonellae and shigellae in Bantu children consuming drinking water of improved quality. *South African Medical Journal* January, 46-9.

Rifkin, S.B. and Walt, G. (1986). Why health improves: defining the issues concerning 'comprehensive Primary Health Care' and 'selective Primary Health Care'. *Social Science Medicine* **23**, 559-66.

Rocuts, F.K. (1985). *Integrated health service at an urban marginal level in Colombia: Santa Fe de Bogota Foundation*. Paper prepared for a workshop on Community health and the urban poor, organized by the London School of Hygiene and Tropical Medicine, OXFAM, and UNICEF, 7-12 July 1985, Oxford, England.

Rodhe, J.E. (1983). Why the other half dies: the science and politics of child mortality in the Third World. *Assignment Children* **61/62**, 35-67.

Rossi-Espagnet, A. (1984). *Primary Health Care in urban areas: reaching the urban poor in developing countries*. A state of the art report by UNICEF and WHO. Report No. 2499M. World Health Organization, Geneva.

Rossi-Espagnet, A. (1985). *Health and the urban poor*. Paper prepared for a workshop on Community health and the urban poor, organized by the London School of Hygiene and Tropical Medicine, OXFAM, and UNICEF, 7-12 July 1985, Oxford, England.

Sabir, N.I. (1984). Why do girls die more? Sex differences in growth and child-rearing practices in a slum area in Lahore. Unpublished M.Sc. thesis. Tropical Child Health Unit, Institute of Child Health, London.

Sakuntanaga, P. (1985). *Community health and the urban poor: Bangkok, Thailand*. Paper prepared for a workshop on Community health and the urban poor, organized by the London School of Hygiene and Tropical Medicine, OXFAM, and UNICEF, 7-12 July 1985, Oxford, England.

Satterthwaite, D.E. (1985). *Health related work of the human settlements programme, International Institute for Environment and Development*. Paper prepared for a workshop on Community health and the urban poor, organized by the London School of Hygiene and Tropical Medicine, OXFAM, and UNICEF, 7-12 July 1985, Oxford, England.

Shah, K.P. (1983). Low birth weight, maternal nutrition and birth spacing. *Assignment Children* **61/62**, 177-91.

Shubert, C. (1986). *Reflections on Primary Health care projects among the urban poor: implications for large-scale programme development*. Paper prepared for the UNICEF/WHO Inter-regional Consultation on Primary Health Care in urban areas, 7-11 July 1986, Manila, Philippines.

Singh, A.M. (1980). Income generation and community development in Hyderabad. *Assignment Children* **49/50**, 173-95.

Stanton, B., Clemens J., Koblinsky, M., and Khair, T. (1985). The urban volunteer programme in Dhaka: a community based Primary Health Care and research initiative. *Tropical and Geographical Medicine* **37**, 183-7.

Stephens, B. (1976). *Gaborone migration study*. National Institute for research in development and African Studies, Gaborone.

Stephens, B., Mason, J.P., and Isely, R.B. (1985). Health and low-cost housing. *World Health Forum* **6**, 59–62.

Sudharto, P. (1983). *Experiences in the field of urban primary health care in Jakarta.* Report of the joint UNICEF/WHO meeting on primary health care in urban areas, 25–29 July 1983, Geneva. WHO document SHS/HSR/83.1.

Sukthankar, D.M. (1983). *Population health problems of the City of Bombay.* Paper prepared for the joint UNICEF/WHO meeting on Primary Health Care in Urban Areas, 25–29 April 1983. WHO document SHS/HS/83.1 World Health Organization, Geneva.

Suliman, A. (1985). *Khartoum peripheral PHC Centres.* A paper prepared for a workshop on Community health and the urban poor, organized by the London School of Hygiene and Tropical Medicine, OXFAM, and UNICEF, 7–12 July 1985, Oxford, England.

Suva, E.G. (1984). *Urban Primary Health Care Department in Manila—from experiment to implementation circa 1983–84.* Paper presented at joint UNICEF/WHO consultation on Primary Health Care in urban areas, 15–19 October, Guayaquil, Ecuador. WHO Report SHS/84.5.

Suva, E.G. (1985). *A lay reporting mechanism for the development of a health information monitoring system in urban Primary Health Care—The Manila Experience.* Paper presented at WHO interegional meeting on Lay reporting in information support to health for all strategy management, 8–15 October 1985, Manila, Philippines.

Tabibzadeh, I. (1985). *Problems of scaling up and enlarging programmes, finance and other support.* Paper prepared for a workshop on Community health and the urban poor, organized by the London School of Hygiene and Tropical Medicine, OXFAM, and UNICEF, 7–12 July 1985, Oxford, England.

Tacon, P. (1981). My child minus one. Unpublished paper. UNICEF, New York.

Tan, J.G. (1985). *Results of the first national consultation on urban Primary Health Care in the Philippines, 16–20 May 1985, Cebu City.* Paper prepared for a workshop on Community health and the urban poor organized by the London School of Hygiene and Tropical Medicine, OXFAM, and UNICEF, 7–12 July 1985, Oxford, England.

Teferra, J. (1985). *A case study of the health component in Kebele 41, Kefetegna 3: community based integrated urban development and organizational problems of scaling up and enlarging the programmes.* Paper prepared for a workshop on Community health and the urban poor, organized by the London School of Hygiene and Tropical Medicine, OXFAM, and UNICEF, 7–12 July 1985, Oxford, England.

Tekce, B. and Shorter, F. (1984). Socio-economic determinants of child mortality and intermediary processes: findings from a study of squatter settlements in Amman. Supplement on child survival, strategies for research (ed. H. Mosley and L. Chen). *Population and Development Review* **10**, 257–80.

Teller, C.H. (1981). *The population dynamics of urbanization and some implications for the health sector.* Paper prepared for the PAHO/AMRO Regional Technical Convention on the development of health services and primary health care in urban areas and big cities, 16–20 November 1981, Washington DC, (original in Spanish).

Theunyck, S. and Dia, M. (1981). The young and the less young in peri-urban area in Mauritania. *African Environment* **14/15/16**.

Turner, J. (1976). *Housing by people: towards autonomy in building environments.* Marion Boyars, London.

UNESCO (United Nations Educational, Scientific and Cultural Organization) (1976). *Working paper: meeting of experts on urban problems and the education of town planners.* Chandigarh, November (mimeo).

Unger, J.P. and Killingsworth, J.R. (1986). Selective Primary Health Care: a critical review of methods and results. *Social Science Medicine* **22**(10), 1001–13.

UNICEF (United Nations Children's Fund) (1982). Urban Basic Services: reaching children and women of the urban poor. E/ICEF/L.1440. UNICEF, New York.

UNICEF, Quito. (1983). Primary Health Care in slum areas of Guayaquil, Ecuador. *Assignment Children* **63/64**, 115–31.

UNICEF (1984*a*). Urban agriculture: meeting basic food needs for the urban poor. *Urban Examples* No. 9. UNICEF, New York.

UNICEF (1984*b*). *Urban basic services: reaching children and women of the urban poor.* UNICEF Occasional Paper Series No. 3. UNICEF, New York.

UNICEF (1985). Programmes for our disabled children. *Urban Examples* No. 11. UNICEF, New York.

UNICEF (1986). Child survival and development and urban basic services. *Urban Examples* No. 12. UNICEF, New York.

United Nations (1970). Quoted in UNICEF (1977). A strategy for basic services. UNICEF, New York.

United Nations (1980). Patterns of urban and rural population growth. *United Nations Population Studies* No. 68. Department of International Economic and Social Affairs, United Nations, New York.

Vaughan, J.P. (1980). Barefoot or professional? Community health workers in the Third World. *Journal of Tropical Medicine and Hygiene* **83**, 3–10.

Verhasselt, Y. (1985) Editorial—urbanization and health in the developing world. *Social Science and Medicine* **21**, 483.

Walt, G. (1984). *Information, education and communication for health.* A background paper for the WHO/UNICEF Joint Committee on Health Policy. WHO, Geneva.

Walt, G. and Constantinides, P. (1984). *Community health education in developing countries.* An historical overview and policy implications, with a selected annotated bibliography. Evaluation and Planning Centre for Health Care, London School of Hygiene and Tropical Medicine, London.

Wayburn, L. (1985). Urban gardens: a lifeline for cities? *The Urban Edge* **9**, 6.

Wegelin, E.A. and Charond, C. (1983). Home improvement, housing, finance and security of tenure in Bangkok slums. In *Land for housing the poor* (ed. Angel, Archer, Tanphiphat and Wegelin). Select Books, Singapore.

Werner, D. (1981). Village health worker: lackey or liberator? *World Health Forum* **2**, 46–54.

WFS (World Fertility Survey) (1984). *World Fertility Survey, major findings and implications*. International Statistical Institute, The Hague.

White, G.F., Bradley, D.J., and White, A.U. (1972). *Drawers of water*. University of Chicago Press, Chicago.

WHO (World Health Organization) (1978). Primary Health Care: Report of the International Conference on Primary Health Care, Alma Ata, USSR, 6–12 September 1978. *Health for All* Series No. 1. WHO, Geneva.

WHO (1981). Global strategy for Health for all by the year 2000. *Health for All Series* No. 3. WHO, Geneva.

WHO (1983*a*). *New approaches to health education in Primary Health Care*. Report of a WHO Expert Committee. Technical Report Series No. 690. WHO, Geneva.

WHO (1983*b*). Report on a joint UNICEF/WHO meeting on Primary Health Care in urban areas, Geneva, 25–29 July 1983. WHO document SHS/HSR/83.1. WHO, Geneva.

WHO (1984*a*). *Mental health care in developing countries: a critical appraisal of research findings*. Report of a WHO study group. WHO Technical Report Series No. 698. WHO, Geneva.

WHO (1984*b*). Joint UNICEF/WHO consultation on Primary Health Care in urban areas, Guayaquil, Ecuador, 15–19 October 1984. WHO Report No. SHS/845. WHO, Geneva.

WHO (1985). *Primary Health Care in urban areas: Reaching the urban poor of developing countries*. Report and recommendations made at 25th session of UNICEF/WHO Joint Committee on Health Policy, Geneva, 28–30 January. Report number JC25/UNICEF—WHO/85.5.

WHO (1986). UNICEF/WHO Inter-regional consultation on Primary Health Care in urban areas, Manila, Philippines, 7–11 July 1986. SHS/IMS/86.1. World Health Organization, Geneva.

WHO (1987). *Hospitals and Health for All*. Report of a WHO Expert Committee on the role of hospitals at the first referral level. Technical Report Series. World Health Organization, Geneva.

World Bank (1975). *Urban sector survey, Manila*. World Bank, Washington DC.

World Bank (1978). The urban poor are also sick. *The Urban Edge* January. World Bank, Washington DC.

World Bank (1984). *Staff appraisal report: Brazil second health project*. Sao Paulo basic health care and National health policy studies. Population, health and nutrition department, World Bank, Washington DC.

World Bank (1984). *World Development Report 1984*. Oxford University Press, New York.

Wray, J.D. (1985). *Nutrition and health in urban slums: an overview*. Paper prepared for a workshop on Community health and the urban poor, organized by the London School of Hygiene and Tropical Medicine, OXFAM, and UNICEF, 7–12 July 1985, Oxford, England.

Yang, T.H. (1982*a*). *Participation of primary health workers in urban malaria/mosquito control programmes*. Paper presented at the WHO S.E. Asia workshop on regional manpower requirements for urban malaria/mosquito control programmes, 30 October–12 November 1981, Hyderabad, India. WHO Report WHO/MAL/82.983.

Yang, T.S. (1982*b*). *Community participation in urban mosquito-borne disease/ mosquito control programmes.* Paper presented at the WHO S.E. Asia workshop on Regional manpower requirements for urban malaria/mosquito control programmes, 30 October–12 November 1981, Hyderabad, India, WHO Report WHO/MAL/82.984.

Yesudian, C.A.K., White, D.K. and Ranken, J.P. (1985). Health for the slums. *World Health Forum* **6**, 257–9.

Youssef, N. and Hetler, C. (1983). Establishing the economic conditions of women headed households in the Third World: a new approach. In *Women and poverty in the Third World* (ed. M. Buvinic, Lycette, M.A. and Mcgreevy W.P. *et al.*). John Hopkins University Press, Baltimore.

Yusof, K. (1982). Sang Kancil: care for urban squatters in Malaysia. *World Health Forum* **3**, 278–81.

Yusof, K. (1985*a*). *Primary Health Care in the urban squatter populations of Kuala Lumpur (Sang Kancil Project).* Paper prepared for a workshop on Community health and the urban poor, organized by the London School of Hygiene and Tropical Medicine, OXFAM, and UNICEF, 7–12 July 1985, Oxford, England.

Yusof, K. (1985*b*). *Demographic and fertility characteristics of four squatter settlements (Sang Kancil Project Evaluation).* Paper prepared for workshop on Community health and the urban poor, organized by the London School of Hygiene and Tropical Medicine, OXFAM, and UNICEF, 7–12 July 1985, Oxford, England.

Zwingmann, C. (1978). *Uprooting and related phenomena—a descriptive bibliography.* WHO publication MNH/78.23. WHO, Geneva.

City index

Author index

Subject index